GUIDELINES FOR ENABLING CONDITIONS AND CONDITIONAL MODIFIERS IN LAYER OF PROTECTION ANALYSIS

This book is one in a series of process safety guideline and concept books published by the Center for Chemical Process Safety (CCPS). Please go to *www.wiley.com/go/ccps* for a full list of titles in this series.

GUIDELINES FOR ENABLING CONDITIONS AND CONDITIONAL MODIFIERS IN LAYER OF PROTECTION ANALYSIS

Center for Chemical Process Safety
New York, NY

WILEY

Library of Congress Cataloging-in-Publication Data:

Guidelines for enabling conditions and conditional modifiers in layer of protection analysis / Center for Chemical Process Safety, New York, NY.
   pages cm.
  Includes bibliographical references and index.
  ISBN 978-1-118-77793-0 (hardback)
   1. Chemical processes—Safety measures. 2. Chemical processes—Safety standards. I. American Institute of Chemical Engineers. Center for Chemical Process Safety.
  TP150.S24G855 2014
  660—dc23   2013020445

Printed in the United States of America.

10 9 8 7 6 5 4 3 2 1

List of Figures

Abbreviations and Acronyms

AEGL	Acute Exposure Guideline Level
AIChE	American Institute of Chemical Engineers
AIHA	American Industrial Hygiene Association
API	American Petroleum Institute
BPCS	Basic process control system
CCPS	AIChE Center for Chemical Process Safety
CPI	Chemical process industry
CPQRA	Chemical Process Quantitative Risk Analysis
DDT	Deflagration-to-detonation transition
DTL	Dangerous toxic load
EPA	U.S. Environmental Protection Agency
ERPG	Emergency Response Planning Guideline
ETA	Event Tree Analysis
FMEA	Failure Modes and Effects Analysis
FMECA	Failure Modes, Effects, and Criticality Analysis
FTA	Fault Tree Analysis
HAZOP	Hazard and Operability [Study]
IDLH	Immediately Dangerous to Life and Health
IPL	Independent protection layer
LC$_{LO}$	Lethal concentration, low
LC$_{50}$	Lethal concentration, 50% mortality
LOPA	Layer of –Protection Analysis

LOPC   Loss of primary containment

MAWP   Maximum allowable working pressure

NFPA   National Fire Protection Association

P   Probability (dimensionless)

PFD   Probability of failure on demand

PSV   Pressure safety valve

QRA   Quantitative risk analysis

RV   Relief valve

SIF   Safety instrumented function

SIS   Safety instrumented system

SLOD   Significant Likelihood of Death

SLOT   Specified Level of Toxicity

U.K.   United Kingdom

U.S.   United States

Glossary

Abnormal situation: A disturbance or series of disturbances in a process that cause plant operations to deviate from their normal state. In the context of hazard evaluation procedures, synonymous with **deviation**.

Administrative control: A procedural requirement for directing and/or checking engineered systems or human performance associated with plant operations.

Auditability: The ability to inspect information, documents, and procedures that demonstrate the adequacy of and adherence to the design, inspection, maintenance, testing, and operation practices used to achieve the other core attributes.

Autoignition temperature: The lowest temperature at which a fuel/oxidant mixture will spontaneously ignite under specified test conditions, with no other sources of ignition present.

Basic process control system (BPCS): A system that responds to input signals from the process and its associated equipment, other programmable systems, and/or from an operator, and generates output signals causing the process and its associated equipment to operate in the desired manner and within normal production limits.

Cause: In the context of hazard evaluation procedures, an **initiating cause**.

Common cause failure: The occurrence of two or more failures that result from a single event or circumstance.

Conditional modifier: One of several possible probabilities included in scenario risk calculations, generally when risk criteria endpoints are expressed in impact terms (e.g., fatalities) instead of in primary loss event terms (e.g., release, vessel rupture). Conditional modifiers include, but are not limited to: probability of a hazardous atmosphere, probability of ignition, probability of explosion, probability of personnel presence, probability of injury or fatality, and probability of equipment damage or other financial impact.

Consequence: Result of a specific event. In the context of qualitative hazard evaluation procedures, the consequences are the effects following from the initiating cause, with the consequence description taken through to the loss event and sometimes to the loss event impacts. In the context of quantitative risk analyses, the consequence refers to the physical effects of the loss event usually involving a fire, explosion, or release of toxic or corrosive material.

Consequence analysis: The analysis of the expected effects of incident outcome cases, independent of frequency or probability.

Contain and control measures: Primary containment system, basic process control system, operating procedures and training, and other measures to keep process materials and energies confined within the primary containment system and to keep the process within safe design and operating limits, thus avoiding abnormal situations and loss-of-containment events that could lead to loss, damage and injury impacts.

CPQRA: The acronym for Chemical Process Quantitative Risk Analysis. It is the process of hazard identification followed by numerical evaluation of incident consequences and frequencies, and their combination into an overall measure of risk when applied to the chemical process industry. It is particularly applied to episodic events. It differs from, but is related to, a Probabilistic Risk Assessment (PRA), a quantitative tool used in the nuclear industry.

Deviation: A process condition outside of established design limits, safe operating limits, or standard operating procedures.

Enabling condition: A condition that is not a failure, error or a protection layer but makes it possible for an incident sequence to proceed to a consequence of concern. It consists of a condition or operating phase that does not directly cause the scenario, but that must be present or active in order for the scenario to proceed to a loss event; expressed as a dimensionless probability.

Enabling event: Another term sometimes used for enabling condition. The term enabling condition is preferred, since enabling conditions are not generally events but rather conditional states.

Endpoint: The furthest extent in an incident sequence to which a LOPA or QRA scenario evaluation is taken in the scenario analysis. Depending on the approach used, the endpoint could be a material/energy release magnitude, a qualitative category of potential loss and harm impacts, an order-of-magnitude impact category with or without conditional modifiers, or fully quantified loss and harm impacts.

Episodic event: An unplanned event of limited duration, usually associated with an incident.

Episodic release: A release of limited duration, usually associated with an incident.

Event: An occurrence involving a process that is caused by equipment performance or human action or by an occurrence external to the process. Events include initiating events, loss events and success or failure of safeguards.

Event sequence: See **Incident sequence.**

Event tree: A logic model that graphically portrays the combinations of events and circumstances in an incident sequence.

External event: Event external to the system caused by (1) a natural hazard F earthquake, flood, tornado, extreme temperature, lightning, etc., (2) a human-induced event F aircraft crash, missile, nearby industrial activity, fire, sabotage, etc., or (3) an interruption of a service such as electric power or process air.

Failure: An unacceptable difference between expected and observed performance.

Failure mode: A symptom or condition by which a failure is observed. A failure mode might be identified as loss of function; premature or spurious function (function without demand); an out-of-tolerance condition; or a physical characteristic such as a leak observed during inspection.

Failure Modes and Effects Analysis (FMEA): A systematic, tabular method for evaluating and documenting the effects of known types of component failures.

Failure Modes, Effects, and Criticality Analysis (FMECA): A variation of FMEA that includes an estimate of the potential severity of consequences of a failure mode.

Fault tree: A logic model that graphically portrays the combinations of failures that can lead to a specific main failure or incident of interest (Top event).

Frequency: The number of occurrences per unit time at which observed events occur or are predicted to occur.

Hazard: A physical condition or chemical presence that has the potential for causing harm to people, property and/or the environment.

Hazard and Operability (HAZOP) Study: A scenario-based hazard evaluation procedure in which a team uses a series of guide words to identify possible deviations from the intended design or operation of a process, then examines the potential consequences of the deviations and the adequacy of existing safeguards.

Hazard evaluation: Identification of individual hazards of a system, determination of the mechanisms by which they could give rise to undesired events, and evaluation of the consequences of these events on health (including public health), environment and property. Uses qualitative techniques to pinpoint weaknesses in the design and operation of facilities that could lead to incidents.

Hazard identification: The inventorying of material, system, process and plant characteristics that can produce undesirable consequences through the occurrence of an incident.

Hazardous event: See **Loss event.**

HAZOP/LOPA Study: The extension of a HAZOP Study to include aspects of a LOPA, including selecting identified scenarios for further analysis; evaluating the initiating event frequency, consequence severity and effectiveness of IPLs on an order-of-magnitude basis; considering enabling conditions and/or conditional modifiers as appropriate when evaluating scenario risk; and comparing the calculated scenario risk to a risk goal to determine the adequacy of existing risk control measures.

Human error: Any human action (or lack thereof) that exceeds some limit of acceptability (i.e., an out-of-tolerance action) where the limits of human performance are defined by the system. Includes actions by designers, operators, or managers that may contribute to or result in incidents.

Impact: A measure of the ultimate loss and harm of a loss event. Impact may be expressed in terms of numbers of injuries and/or fatalities, extent of environmental damage and/or magnitude of losses such as property damage, material loss, lost production, market share loss, and recovery costs.

Incident: An unplanned event or sequence of events that either resulted in or had the potential to result in adverse impacts.

Incident sequence: A series of events composed of an initiating event and intermediate events leading to an undesirable outcome.

Independent protection layer (IPL): A device, system, or action that is capable of preventing a scenario from proceeding to the undesired consequence regardless of the initiating event or the action of any other protection layer associated with the scenario.

Initiating cause: In the context of hazard evaluation procedures, the operational error, mechanical failure, or external event or agency that is the first event in an incident sequence and marks the transition from a normal situation to an abnormal situation.

Initiating event: The minimum combination of failures or errors necessary to start the propagation of an incident sequence. It can be comprised of a single initiating cause, multiple causes, or initiating causes in the presence of enabling conditions. (The term initiating event is the usual term employed in Layer of Protection Analysis to denote an initiating cause or, where appropriate, an aggregation of initiating causes with the same immediate effect, such as "BPCS failure resulting in high reactant flow." See Appendix A for a discussion of the rare situation where an initiating event consists of two different initiating causes occurring at the same time.)

Intermediate event: An event that occurs after the initiating event and before the loss event in an incident sequence.

Layer of protection: A device, system, or action, supported by a management system, that is capable of preventing an initiating event from propagating to a specific loss event or impact.

Layer of Protection Analysis (LOPA): An approach that analyzes one incident scenario (cause-consequence pair) at a time, using predefined values for the initiating event frequency, independent protection layer failure probabilities, and consequence severity, in order to compare a scenario risk estimate to risk criteria for determining where additional risk reduction or more detailed analysis is needed. Scenarios are identified elsewhere, typically using a scenario-based hazard evaluation procedure such as a HAZOP Study.

Likelihood: A measure of the expected probability or frequency of occurrence of an event. This may be expressed as an event frequency (e.g., events per year), a probability of occurrence during a time interval (e.g., annual probability) or a conditional probability (e.g., probability of occurrence, given that a precursor event has occurred).

Loss event: Point in time in an abnormal situation when an irreversible physical event occurs that has the potential for loss and harm impacts. Examples include release of a hazardous material, ignition of flammable vapors or ignitable dust cloud, and overpressurization rupture of a tank or vessel. An incident might involve more than one loss event, such as a flammable liquid spill (first loss event) followed by ignition of a flash fire and pool fire (second loss event) that heats up an adjacent vessel and its contents to the point of rupture (third loss event). Generally synonymous with **hazardous event**.

Loss of primary containment (LOPC): An unplanned or uncontrolled release of material from primary containment, including non-toxic and non-flammable materials (e.g., steam, hot condensate, nitrogen, compressed CO_2 or compressed air).

Mitigate: Reduce the impact of a loss event.

Mitigative safeguard: A safeguard that is designed to reduce loss event impact.

Operator: An individual responsible for monitoring, controlling, and performing tasks as necessary to accomplish the productive activities of a system. Often used in a generic sense to include people who perform all kinds of tasks (e.g., reading, calibration, maintenance).

Process safety management: A program or activity involving the application of management principles and analytical techniques to ensure the safety of process facilities. Sometimes called **process hazard management**.

Preventive safeguard: A safeguard that forestalls the occurrence of a particular loss event, given that an initiating cause has occurred; i.e., a safeguard that intervenes between an initiating cause and a loss event in an incident sequence. (Note that containment and control measures are also preventive in the sense of preventing initiating causes from occurring; however, the term preventive safeguard in the context of hazard evaluation procedures is used with the specific meaning given here.)

Primary containment: A tank, vessel, pipe, transport vessel or equipment intended to serve as the primary container for, or used for the transfer of, a material. Primary containers may be designed with secondary containment systems to contain or control a release from the primary containment.   Secondary containment systems include, but are not limited to, tank dikes, curbing around process equipment, drainage collection systems into segregated oily drain systems, the outer wall of double-walled tanks, etc.

Probability: The expression for the likelihood of occurrence of an event or an event sequence during an interval of time, or the likelihood of success or failure of an event on test or on demand.   Probability is expressed as a dimensionless number ranging from 0 to 1.

Probability of failure on demand (PFD):   The probability that a system will fail to perform a specified function on demand (i.e., when challenged or needed).

Probit: A random variable with a mean of 5 and a variance of 1, which is used in various effect models.

Quantitative risk analysis (QRA):   The systematic development of numerical estimates of the expected frequency and severity of potential incidents associated with a facility or operation based on engineering evaluation and mathematical techniques.

Revealed failure: A failure that may be immediately or almost immediately apparent through an alarm or indicator system. This can lead to corrective action within a relatively short period of time.

Risk: The combination of the expected frequency (events/year) and severity (effects/event) of a single potential incident or group of incidents.

Risk analysis: The estimation of scenario, process, facility and/or organizational risk by identifying potential incident scenarios, then evaluating and combining the expected frequency and impact of each scenario having a consequence of concern, then summing the scenario risks if necessary to obtain the total risk estimate for the level at which the risk analysis is being performed.

Risk assessment: The process by which the results of a risk analysis (i.e., risk estimates) are used to make decisions, either through relative ranking of risk reduction strategies or through comparison with risk targets.

Risk control measure: Characteristic associated with a process that is expected to reduce the likelihood and/or severity of a loss event.

Risk management: The systematic application of management policies, procedures, and practices to the tasks of analyzing, assessing, and controlling risk in order to protect employees, the general public, the environment, and company assets, while avoiding business interruptions. Includes decisions to use suitable engineering and administrative controls for reducing risk.

Safeguard: Any device, system, or action that would likely interrupt the chain of events following an initiating cause or that would mitigate loss event impacts. See **Preventive safeguard**; **Mitigative safeguard**. See also **Contain and control measures**.

Safety system: Equipment and/or procedures designed to limit or terminate an incident sequence, thus avoiding a loss event or mitigating its consequences.

Scenario: An unplanned event or incident sequence that results in a loss event and its associated impacts, including the success or failure of safeguards involved in the incident sequence.

Source term: The release parameters (e.g. magnitude, rate, duration, orientation, temperature) that are the initial conditions for determining the consequences of the loss event for a hazardous material and/or energy release to the surroundings. For vapor dispersion modeling, it is the estimation, based on the release specification, of the actual cloud conditions of temperature, aerosol content, density, size, velocity and mass to be input into the dispersion model.

Top event: The loss event or other undesired event at the "top" of a fault tree that is traced downward to more basic failures using Boolean logic gates to determine its possible causes.

Unrevealed failure: A failure that may lie dormant in the system and only be discovered as a result of a thorough diagnostic testing procedure.

What-If Analysis: A scenario-based hazard evaluation procedure using a brainstorming approach in which typically a team that includes one or more persons familiar with the subject process asks questions or voices concerns about what could go wrong, what consequences could ensue, and whether the existing safeguards are adequate.

What-If/Checklist Analysis: A What-If Analysis that uses some form of checklist or other listing of broad categories of concern to structure the what-if questioning.

Acknowledgements

The Center for Chemical Process Safety (CCPS) thanks all of the members of the Enabling Conditions and Conditional Modifiers Subcommittee of CCPS' Technical Steering Committee for providing input, reviews, technical guidance and encouragement to the project team throughout the preparation of this book. CCPS also expresses appreciation to the members of the Technical Steering Committee for their advice and support.

The CCPS staff liaison for this project was John F. Murphy, who also coordinated meetings and facilitated subcommittee reviews and communications. The subcommittee had the following members whose significant efforts and contributions are gratefully acknowledged:

Wayne Chastain, Eastman Chemical Company, Subcommittee Chair

Stanley A. Urbanik, Du Pont, Subcommittee Co-Chair

Larry Bowler, SABIC

Bill Bridges, Process Improvement Institute, Inc.

Andrew Carpenter, Exponent, Inc.

Chris Devlin, Celanese Chemicals

Thomas Dileo, Albermarle

Jeffrey Fox, Dow Corning Corp.

Randy Freeman, S&PP Consulting

Michela Gentile, BP

Kieran Glynn, BP

Kenneth Harrington, Chevron Phillips Chemical Company

David Kahn, AcuTech Consulting

Kimberly Mullins, Chevron Phillips Chemical Company

Jack Reisdorf, Fluor Enterprises

Kathy Shell, AE Solutions

Bob Stack, The Dow Chemical Company

Angela Summers, SIS-Tech Solutions

Dave Thompson, Flint Hills Resources

Vincent Van Brunt, University of South Carolina

Robert Wasileski, NOVA Chemicals Corp.

Tim Wagner, The Dow Chemical Company

Unwin Company (Columbus, Ohio) prepared these *Guidelines* under contract to the Center for Chemical Process Safety. Robert W. Johnson was Unwin Company's project manager and lead author and Steven W. Rudy was co-author. Julie Nelson assisted with text editing.

As is standard practice for CCPS projects, in order to achieve guidelines with broad input and technical accuracy, opportunity was given for knowledgeable persons to peer review these *Guidelines* before publication. The following persons peer reviewed this document:

John Alderman, Hazard & Risk Analysis
Peter Clarke, exida Asia Pacific
Art Dowell, A M Dowell III PE PLLC
Bob Gale, Emerson Process Management
Peter N. Lodal, Eastman Chemical Company
Philip M. Myers, Advantage Risk Solutions, Inc.
Brad Newman, Praxair
Ronald Nichols, aeSolutions
Keith R. Pace, Praxair
Adrian Sepeda, CCPS Staff Consultant
Hal Thomas, exida
Nora Williams, Flint Hills Resources

CCPS and the Unwin Company project team gratefully acknowledge the valuable suggestions and feedback submitted by the peer reviewers.

Preface

The American Institute of Chemical Engineers (AIChE) has been closely involved with process safety and loss control issues in the chemical and allied industries for more than four decades. Through its strong ties with process designers, constructors, operators, safety professionals, and members of academia, AIChE has enhanced communication and fostered continuous improvement of the industry's high safety standards. AIChE publications and symposia have become information resources for those devoted to understanding the causes of incidents and discovering better means of preventing their occurrence and mitigating their consequences.

The Center for Chemical Process Safety (CCPS) was established in 1985 by AIChE to develop and disseminate technical information for use in the prevention of major chemical incidents. CCPS is supported by over 100 sponsoring companies in the chemical process industry (CPI) and allied industries. These companies provide the necessary funding and professional experience for its technical subcommittees.

CCPS' first project was the preparation of *Guidelines for Hazard Evaluation Procedures*. CCPS achieved its stated goal with the publication of the *Guidelines* in 1985, and has since continued to foster the development of process safety knowledge in all industries.

Layer of Protection Analysis (LOPA) is a tool for analyzing and assessing scenario risk. LOPA has grown in popularity in the time since the publication of the first CCPS/AIChE guidebook on the subject (*Layer of Protection Analysis: Simplified Process Risk Assessment*; CCPS 2001). It uses estimates of cause frequency, independent protection layer failure probabilities and consequence severity, employing conservative rules for making and combining these estimates.

These *Guidelines for Enabling Conditions and Conditional Modifiers* contain information useful to the experienced analyst and accomplished practitioner.

This book is intended for:

- <u>Managers and engineers responsible for maintaining or establishing LOPA protocols for an operating company or contracting firm.</u> A protocol document can help to establish parameters around the use of enabling conditions and conditional modifiers in a company to ensure consistent use (or non-use) across the company, consistency with the company's risk criteria and avoidance of their over-use. In addition, managers and risk analysts who are establishing or revising corporate risk criteria for single-scenario risk evaluations will find guidance on the interaction between conditional modifiers and risk criteria.

- <u>LOPA practitioners.</u> Guidance is given to help both understand and properly use enabling conditions and conditional modifiers, including knowing when their use is inappropriate.

- <u>Process and process control engineers, operations and maintenance personnel and others who participate in LOPAs.</u> Understanding the contents of this book will help LOPA participants to make appropriate use of enabling conditions and conditional modifiers.

The following describes the organization of this document.

Chapter 1 – Context

- Presents a brief overview of Layer of Protection Analysis (LOPA) and its variations, and summarizes terminology used for evaluating scenarios in the context of a typical incident sequence
- Outlines important theoretical and practical limitations of enabling conditions and conditional modifiers and summarizes what practitioners and managers can reasonably expect from the use of more detailed LOPA and QRA approaches.

Chapter 2 – LOPA Enabling Conditions

- Defines *enabling conditions* and shows how they interrelate with initiating causes
- Defines and illustrates the most common types of enabling conditions
- Discusses the documentation and validation of enabling conditions in the context of a LOPA study.

Chapter 3 – LOPA Conditional Modifiers

- Defines *conditional modifiers* and shows how they interrelate with risk criteria

- Defines and illustrates the most common types of conditional modifiers
- Discusses the documentation, management and validation of conditional modifiers in the context of a LOPA study.

Chapter 4 – Application to Other Methods

- Discusses how enabling conditions and conditional modifiers are used in quantitative risk analyses.
- Illustrates the use of enabling conditions and conditional modifiers when performing HAZOP/LOPA Studies and other similar techniques.
- Discusses the possible application of enabling conditions and/or conditional modifiers in barrier analysis applications.

Appendices. The Appendices provide:

- A discussion of the very limited number of situations in which consideration of simultaneous failures is pertinent, and shows how to quantify them.
- A discussion of peak risk as it applies to the application of enabling conditions and conditional modifiers.
- An example of the kind of rule set an organization might use if it decides to employ enabling conditions and/or conditional modifiers in its LOPAs.

As is true for other CCPS books, these *Guidelines* do not contain a complete program for managing the risk of chemical operations, nor do they give specific advice on how to establish a risk analysis program for a facility or an organization. However, they do provide some of the insights that should be considered when performing more detailed, scenario-based risk evaluations.

These *Guidelines* cannot replace hazard evaluation experience. This book should be used as an aid for the further education of hazard analysts and as reference material for experienced practitioners. Only through both study and experience will hazard analysts become skilled in the use of enabling conditions and conditional modifiers. Using these *Guidelines* within the framework of a complete process safety program can help organizations continually improve the safety, environmental performance and loss prevention in their facilities and operations.

1
Context

The *Guidelines* in this book characterize when and how to apply enabling conditions and conditional modifiers to Layer of Protection Analyses (LOPAs). A LOPA may have consequences and risk criteria expressed in final endpoint (impact) terms such as fatalities or environmental damage, and include *conditional modifiers* such as probability of fatality associated with a material or energy release. It may also take into account probabilities called *enabling conditions* that sometimes apply to scenario initiating events. One way to differentiate these two factors is that <u>enabling conditions</u> are associated with the part of an incident sequence <u>leading up to</u> a release of hazardous material or energy, whereas <u>conditional modifiers</u> are probabilities generally associated with the <u>post-release</u> part of an incident sequence.

As discussed in Chapter 4, these enabling conditions and conditional modifiers may be used with other scenario risk analysis methods such as Quantitative Risk Analyses and HAZOP/LOPA Studies. However, the main focus in this text will be their use in the context of LOPAs.

An overview of Layer of Protection Analysis is given in Section 1.1, followed by some pertinent variations in Section 1.2. Section 1.3 discusses the appropriate application of enabling conditions and conditional modifiers. These decisions on use of conditional modifiers are closely related to a company's risk criteria *endpoints* used when determining loss event impacts, as discussed in Section 1.4.

1.1   LOPA Overview

This section, taken in large part from Center for Chemical Process Safety's *Guidelines for Hazard Evaluation Procedures, Third Edition* (CCPS 2008)[1], provides a brief summary of the methodology for conducting Layer of Protection Analyses as described in the CCPS Concept Book *Layer of Protection Analysis: Simplified Process Risk Assessment* (CCPS 2001), with minor updates. Experienced LOPA practitioners and users may want to advance to Section 1.2, Pertinent LOPA Variations.

Layer of Protection Analysis is a simplified form of quantitative risk analysis that uses order-of-magnitude categories for initiating event frequency, consequence severity and probability of failure of independent protection layers (IPLs) to analyze and assess the risk of one or more incident scenarios. LOPA can

---

[1] See References list at the end of this document for all cited references.

be useful in the process development, process design, operational, maintenance, modification and decommissioning life cycle phases.

Purpose of LOPA

LOPA was developed to help answer questions such as:

- What layers of protection are needed to meet our risk goals?
- How much risk reduction does each layer provide or need to provide?

LOPA can help answer these questions with less time and effort than a full quantitative risk analysis (QRA), although there are instances when use of a complete QRA may be warranted.

LOPA is an order-of-magnitude type of quantitative method (sometimes termed "semi-quantitative") that builds on qualitative hazard evaluations such as HAZOP Studies. By analyzing selected scenarios in detail, effective application of LOPA can determine whether the risk posed by each analyzed scenario has been reduced to meet a specified risk goal. However, if the analyst or team can make a reasonable risk decision using only qualitative methods, then LOPA may not be warranted. Qualitative hazard evaluation methods such as HAZOP Studies are intended to identify a comprehensive set of incident scenarios and qualitatively analyze those scenarios for the adequacy of safeguards. LOPA is generally used to analyze a subset of incident scenarios.

At times, simple order-of-magnitude LOPAs can be expanded with greater rigor by implementing aspects of a more complete QRA. The use of enabling conditions and conditional modifiers falls into this realm.

Description

LOPA is typically applied after, and builds upon, the information gathered in a qualitative hazard evaluation, but can be applied to scenarios gathered from any source, such as an audit or incident investigation. LOPA, in turn, can be used as a screening tool for scenarios prior to application of a full quantitative risk analysis. If desired, LOPA can also be implemented in conjunction with a qualitative scenario-based hazard evaluation method such as a HAZOP Study (see Section 4.2).

After the scope of the study is defined, LOPA consists of the six steps summarized below. The LOPA results can then be used to make risk-based decisions.

Step 1: Identify the scenario screening criteria. Since LOPA typically evaluates scenarios that have been developed in a prior study, a first step by the LOPA analyst or team is to screen these scenarios, and the most common screening method is based on consequence. Other screening criteria may also be employed, such as based on "unmitigated risk" (consequence severity combined with initiating cause frequency estimates) or based on the judgment of the hazard review team as to which scenarios warrant closer examination.

LOPA consequence severity estimates and consequence screening thresholds may be defined in a number of ways, each having strengths and weaknesses and varying in the degree of conservatism incorporated into the analysis, with simpler methods generally being more conservative:

- Method 1 – Category approach without direct reference to human harm. Consequences are categorized in terms of the type and magnitude of a release or other consequence characteristic, rather than explicitly defining the final consequence in terms of the number and severity of injuries that may result from a particular release.
- Method 2 – Qualitative estimates of human harm. Human impacts are considered, usually allowing direct comparison with organizational guidelines, but estimates of impact magnitudes are arrived at using qualitative judgment. This method does not explicitly use conditional modifiers such as probability of personnel presence.
- Method 3 – Qualitative or order-of-magnitude estimates of human harm, with adjustments for post-release probabilities (conditional modifiers). This method extends the qualitative judgment in Method 2 using additional probabilities, giving a more quantitative (order-of-magnitude) estimate of the severity of human harm.
- Method 4 – Quantitative estimates of human harm. This method is similar to Method 3 but uses detailed analyses in determining the effects of a release and its effects upon individuals and equipment. Tools associated with full Chemical Process Quantitative Risk Analyses (CPQRAs) may be used here, including dispersion and blast effects analysis. The use of these tools adds complexity and time as well as a need for expertise and other resources such as computational tools. The level of sophistication required for these consequence analyses may be disproportionate to the order-of-magnitude frequency estimates employed within LOPA.

Step 2: Select an incident scenario. LOPA is applied to one scenario at a time. The scenario can come from other analyses, such as qualitative hazard evaluations and/or management-of-change reviews, but each scenario must describe a single initiating event – loss event ("cause-consequence") pair, except for an uncommon situation where a scenario requires two concurrent initiating events. (This situation, which may warrant using quantitative risk analysis, is discussed in Appendix A.)

When scenarios are selected from a qualitative hazard evaluation, such as from a HAZOP Study, they may need to be separated into multiple scenarios for evaluation. For example, where a scenario involves an emergency relief system, the LOPA could be applied both to the case in which the emergency relief device does not function properly (usually involving a greater severity but lesser likelihood) and to the case in which it does function properly (usually involving a lesser severity but greater likelihood).

Step 3: Identify the initiating event of the scenario and determine its frequency. The scenario *initiating event* must lead to the loss event, given failure of all of the preventive safeguards. (An initiating event is comprised of one or more aggregated initiating causes, such as from a HAZOP Study. Initiating causes can only be aggregated if they are protected by the same IPLs and lead to the same loss event.) Most companies provide guidance on estimating the frequency to achieve consistency in LOPA results; suggested frequencies are also given by CCPS (2001). The team should determine whether the suggested value is appropriate, based on plant historical performance and/or experience with the initiating event occurring under similar plant conditions. Other factors may also enter into the determination of the initiating event frequency, such as extraordinary design or maintenance or a keylock system to make a human error less likely.

Background aspects, such as the probability that the process is in a certain mode of operation at the time another failure occurs, are not initiating events but *enabling conditions*. In LOPA, their probabilities modify the initiating event frequency. These aspects are the subject of Chapter 2 of these *Guidelines*.

Step 4: Identify the IPLs and estimate the PFD of each IPL. The heart of the LOPA methodology is recognizing the existing preventive safeguards that meet the requirements of *independent protection layers* (IPLs) for a given scenario. (All IPLs are safeguards, but not every safeguard meets the requirements of being an IPL.)

An IPL is a device, system, or action that is capable of keeping a scenario from proceeding to the undesired loss event, independent of the initiating event or the action of any other layer of protection associated with the scenario. A preventive safeguard meets the requirements of being an IPL when it is designed and managed to achieve the following seven core attributes:

- *Independent* – the performance of a protection layer not being affected by the initiating event or by the actions of other protection layers.
- *Functional* – capable of operating successfully in response to a specific abnormal condition.
- *Integrity* – the risk reduction that can reasonably be expected given the protection layer's design and management.
- *Reliable* – assurance that a protection layer will operate as intended under stated conditions for a specified time period.
- *Validated, maintained, and audited* – implementing, maintaining and verifying information, documents, and procedures that demonstrate the adequacy of and adherence to the design, inspection, maintenance, testing, and operation practices used to achieve the other core attributes.

- *Access security* – the use of administrative controls and physical means to reduce the potential for unintentional or unauthorized changes.
- *Management of change* – the formal process used to review, document, and approve modifications that are not replacements in kind, prior to implementation of the modifications.

IPLs can be viewed as "lines of defense" against potential incident scenarios. The independence of the IPL from the initiating event and from other IPLs is very important. The LOPA team must assess the independence of each IPL and estimate its probability of failure based on the IPL design and management. All IPL equipment should be included in the facility's mechanical integrity program and be subjected to inspection and proof tests as necessary to maintain the target probability of failure on demand (PFD). IPLs depending on operating personnel should have key steps explicitly described in written procedures, with identified personnel being trained and tested on the procedures to show they are capable of responding in time. Access to IPL equipment should be controlled, and proposed changes should undergo management of change review prior to implementation.

One type of IPL is the *safety instrumented system (SIS)*. CCPS has published guidelines covering the life cycle of safety instrumented systems and other instrumented protective systems (CCPS 2007).

The integrity of each IPL is quantified in terms of its probability of failure on demand, which is a dimensionless number between 0 and 1, inclusive. The PFD of an IPL is the probability that, when needed for the scenario in question, the IPL will not perform the required task. Most companies provide a predetermined set of PFD values for use by the LOPA analyst, so the analyst may pick the IPL that best fits the scenario being analyzed and the PFD that best fits the equipment configuration and the facility's integrity management plan.

Step 5:   Calculate the scenario frequency. The overall predicted frequency of realizing the scenario consequence is estimated by mathematically combining the initiating event frequency and the IPL PFDs. Combining methods include arithmetic formulae and graphical approaches. Regardless of the method, most organizations provide a standard form for documenting LOPA intermediate and final results. The following mathematical approach is applicable for low-demand situations (F_i^I less than twice the test frequency for the first IPL; see CCPS 2001 Appendix F).

$$f_i^C = F_i^I \cdot \prod PFD_{ij}$$

$$= F_i^I \cdot PFD_{i1} \cdot PFD_{i2} \cdot \ldots \cdot PFD_{ij}$$

where

f_i^C is the scenario frequency for consequence C for initiating event i

F_i^I is the initiating event frequency for consequence C for initiating event i

PFD_{ij} is the probability of failure on demand (fail-dangerous failure mode) of the j th IPL that protects against consequence C for initiating event i.

For example, if the frequency for the first initiating event (i =1) is estimated to be once every ten years ($F_1^I = 10^{-1}$/year), and two IPLs each having a PFD of 0.01 ($PFD_{11} = 10^{-2}$ and $PFD_{12} = 10^{-2}$) are protecting against this particular initiating event and will prevent consequence C from being realized if either IPL works successfully in response to initiating event 1, then the calculated frequency for consequence C for initiating event 1 (f_1^C) is equal to 10^{-1}/year x 10^{-2} x 10^{-2} = 10^{-5}/year.

An *enabling condition* factor (as described in Chapter 2) may be included in the scenario frequency calculation if an organization chooses to use enabling conditions and if the factor is pertinent to the scenario being evaluated. The enabling conditional probability is combined into the frequency equation above in the same way PFDs are included.

Conditional modifier factors (as described in Chapter 3) may be included in the scenario frequency calculation, depending on the approach used for estimating consequence severity (if Method 3 or 4 is used, as discussed in Step 1 of this Section) and an organization chooses to use conditional modifiers. For example, when a company is estimating the frequency of direct human harm, the analysis may include additional modifying probabilities in the scenario frequency calculation, such as probability of ignition or presence of personnel near a release.   Those probabilities are then included as additional factors in the frequency equation, as in

$$f_i^C = f_i^I * \prod PFD_{ij} * P^{ignition}$$

for the frequency of ignition, and

$$f_i^C = f_i^I * \prod PFD_{ij} * P^{ignition} * P^{person\_present}$$

for the frequency that a person might be present near the fire, and so on.

Step 6: Evaluate the risk to reach a decision concerning the scenario. The frequency of the outcome of interest is multiplied by a factor related to the magnitude of the consequences to obtain the scenario risk:

$$R_k^C = f_k^C * C_k$$

where

R_k^C is the risk of incident outcome of interest k, expressed as a magnitude of consequences per unit time, such as fatalities per year.

f_k^C is the frequency of the outcome of interest k, in inverse time units

C_k is a specific measurement of the severity of consequences (impact) of the incident outcome(s) of interest k (e.g., fatalities, serious injuries, public impacts, environmental impacts, economic loss).

The scenario risk or frequency may be calculated and compared to a specific target, or may be shown on a matrix of consequence versus frequency.

Step 7: Using LOPA to make risk decisions. Once LOPA has been applied to yield order-of-magnitude risk estimates for a scenario, a risk-based decision can be made. This evaluation is normally in relation to an organization's risk criteria related to scenario risks (CCPS 2009). If the calculated risk exceeds the tolerable risk level, the proportion by which the calculated risk exceeds the tolerable risk level indicates by how much the risk must be reduced. Risk reduction can be achieved by various means, including eliminating scenarios by inherent safety approaches, reducing initiating event frequencies, increasing integrity of IPLs, adding more IPLs and/or reducing loss event impacts. Many other uses have been found for LOPA results, including adjusting mechanical integrity programs to emphasize oversight of particular equipment components.

Anticipated Work Product

Implementation of LOPA results in a set of order-of-magnitude risk estimates for the scenarios selected for evaluation. LOPA also includes an assessment of the adequacy of the independent protection layers for each scenario in the form of risk-based decisions and, if deemed necessary, recommendations and specifications for additional IPLs.

An example LOPA worksheet that illustrates these results for a single scenario is shown in Table 1.1. CCPS (2001) can be consulted to understand the details of this example. This example includes one enabling condition (Probability that reactor in condition where runaway reaction can occur on loss of cooling) and two conditional modifiers (Probability of personnel in affected area; Probability of fatal injury) although no risk-reduction credit was given for either of the conditional modifiers (i.e., probability = 1 for each).

Table 1.1 Example LOPA worksheet (adapted from CCPS 2001, Table B.2)

Scenario # 1	**Scenario title:** Cooling water failure results in runaway reaction with potential for reactor overpressure, leakage, rupture, injuries and fatalities. Agitation assumed.		
Date: mm/dd/yyyy	Description	Frequency (per year)	Probability
Consequence description and impact category	Runaway reaction and potential for reactor overpressure, leakage, rupture, injuries, and fatalities **Category 5**		
Risk tolerance criteria (for given Category)	Unacceptable (Greater than) Tolerable (Less than or equal to)	1×10^{-4} 1×10^{-6}	
Initiating event	Loss of cooling water	1×10^{-1}	
Enabling condition	Probability that reactor in condition where runaway reaction can occur on loss of cooling (annual basis)		0.5 (per reactor)
Conditional modifiers (if applicable)	Probability of ignition		N/A
	Probability of personnel in affected area		1
	Probability of fatal injury		1
	Others		N/A
Frequency of consequence without IPLs		5×10^{-2}	
Independent protection layers (IPLs)			
BPCS alarm and human action	Shortstop addition on BPCS loop high reactor temperature alarm; adequate time to respond		1×10^{-1}
Pressure relief valves	With required modifications to system (see Actions; PFD may be conservative if modifications added)		1×10^{-2}
Safety instrumented function (SIF)	SIF to open vent valves to be added (see Actions for design details)		1×10^{-2}
Safeguards (non-IPLs)	**Operator action.** Other operator actions not independent of the same operator already credited. **Emergency cooling system** (steam turbine). Not credited as an IPL as too many common elements (piping, valves, jacket, etc.) that could have initiated initial cooling water failure.		
Total PFD for all IPLs			1×10^{-5}
Frequency of consequence with IPLs		5×10^{-7}	
Risk tolerance criteria met? (Yes/No): Yes, after implementation of actions listed below.			
Actions required to meet risk tolerance criteria	Add SIS for all 3 reactors. Install SIF with minimum PFD = 1×10^{-2} for opening vent valves on high temperature. Separate nozzles and piping for each vent valve. Install separate nozzle and vent lines for each PSV to minimize blockage and common cause. Consider nitrogen purges under all vent valves / PSVs. **Responsible group / person / date:** Plant Technical / J. Doe / mm/yyyy		
Notes	1. Ensure operator response to high temperature meets requirements for IPL. 2. Ensure RV design, installation, maintenance meet requirements for PFD 1×10^{-2} as a minimum.		

1.2 Pertinent LOPA Variations

Users have developed many variations on the basic LOPA methodology summarized in Section 1.1. The variations that are pertinent to the use of enabling conditions and conditional modifiers are discussed in this section. These particular variations are a function of three main factors:

- The resolution of the numerical values used in the LOPA calculations.
- The means by which these values are determined.
- The extent to which loss event consequences are evaluated.

Variations in LOPA Value Resolution

LOPAs most commonly employ frequency and probability estimates with only order-of-magnitude resolution. For example, initiating event frequencies are generally restricted to ten-fold differences from an annual frequency (10/yr, 1/yr, 0.1/yr, 0.01/yr, etc.), and probabilities to 1, 0.1, 0.01, 0.001, etc. Likewise, consequence estimates are typically made on roughly an order-of-magnitude basis.

One variation on this approach is to not limit values to order-of-magnitude categories. The preceding Table 1.1 shows an example of this variation, where an enabling condition probability of 0.5 is included in the analysis. Another variation is to use a resolution of half orders of magnitude. The use of LOPA values that are not necessarily on a full order-of-magnitude basis can facilitate the inclusion of factors such as enabling conditions and conditional modifiers, if the input data supports the use of such factors at greater than an order-of-magnitude resolution.

Significant figures. For most LOPA calculations employing the variation that does not restrict LOPA values to order-of-magnitude categories, it is appropriate to use only one significant figure in the presentation of the LOPA entries and the ensuing calculated values. This needs to be emphasized, since the use of computerized approaches for documenting LOPAs sometimes leads to values being presented with a misleading number of significant figures. By comparison, fully quantitative risk analyses often employ two significant figures in the presentation of calculated risk estimates, although this is often not warranted by the degree of uncertainty of many input values. In this document, the determination and use of enabling condition and conditional modifier probabilities will be discussed from both an order-of-magnitude perspective and from a more quantitative perspective that will allow for their understanding and usage both in LOPAs that do not restrict values to order-of-magnitude categories and in quantitative risk analyses.

Resolution and accuracy of enabling condition and conditional modifier values. The resolution of *enabling condition* and *conditional modifier* probability estimates (as defined and illustrated in Chapters 2 and 3, respectively) will vary considerably, based on the specific factor being evaluated. Order-of-magnitude categories are appropriate for some factors such as probability of ignition that may have a relatively high degree of uncertainty.

Other factors will have historical data capable of supporting values with a much greater degree of resolution. For example, an enabling condition might be that the ambient temperature must be below freezing for a cooling water supply line to freeze upon failure of its heat tracing. Meteorological data for the location of interest is likely available that will give the fraction of the year with ambient temperature below freezing with a fairly high degree of resolution. Other examples of enabling condition and conditional modifier probabilities that may be able to be estimated with a resolution greater than an order-of-magnitude basis are the fraction of time a process unit is in a particular operating mode (such as full recycle or high-fire) and the fraction of time the wind is blowing towards a certain receptor location.

Nevertheless, error bands around some enabling condition and conditional modifier probabilities may be broader than those around typical initiating event frequencies and independent protection layer PFDs. The potential may exist for increasing the overall uncertainty of a LOPA or QRA by employing enabling conditions and/or conditional modifiers, versus not including them in the analysis, just by virtue of having a greater number of factors in the risk equation with each factor having an associated degree of uncertainty.

Where uncertainty does exist for a given factor, several possible approaches are available to those performing a LOPA study or a QRA. These include:

- Selecting a worst-case, most-conservative value
- Selecting a reasonable, conservative value
- Selecting a best-estimate value
- If the uncertainty is too high, not including the enabling condition or conditional modifier in the scenario analysis or QRA.

A suggested approach is to use best-estimate values where data, experience and/or standardization support them. Then, when a particular factor has some uncertainty associated with its frequency or probability, use a reasonable, conservative value for the factor. A check can be made on overall risk estimates that appear unexpectedly low or high by performing a sensitivity analysis on the factors with the greatest degree of uncertainty, and observing the effect of varying their values within their respective possible ranges. This approach avoids the pitfall of combining multiple factors each of which is skewed toward a conservative estimate (i.e., tending toward a higher overall scenario frequency), and ending up with a scenario frequency that is unrealistically high. The above discussion is especially pertinent when employing conditional modifiers, since a typical LOPA scenario evaluation considers only one enabling condition (if any), whereas three or more conditional modifiers might be included in the analysis, each with its own uncertainty.

In summary, one common pitfall in performing LOPAs is to document the individual factors and/or present the results with more significant figures than is warranted by the degree of resolution of the underlying data. Analyses using order-of-magnitude estimates for each factor should also present the final results as an order-of-magnitude value. Analyses using a greater degree of resolution should document individual factors and present the final results with only one significant figure (or as an order of magnitude, if any of the factors are only estimated on an order-of-magnitude basis). Using too many significant figures is misleading regarding the degree of resolution of the risk estimates.

Variations in LOPA Frequency and PFD Determinations

Companies or facilities employing LOPA often have standardized values that are used for various initiating event frequencies (e.g., maintained pump in standard service fails off 0.1/yr) and IPL failure probabilities (e.g., screening-value PFD = 0.01 for a relief valve failing to respond in non-plugging service). The same approach of picking from a standardized list can be used for enabling condition and conditional modifier probabilities. The standardized list should provide criteria for applicability of the factors to the actual scenario being evaluated.

However, it is possible to employ enabling condition and conditional modifier probabilities that are based on a more in-depth analysis of the specific scenario being evaluated. This becomes closer to a QRA approach, where some logic modeling or data analysis may be required to determine the best-estimate probability. The approach used is determined by what will best support the objectives of the LOPA study, consistent with the method and risk criteria established by the operating company.

Variations in LOPA Consequence Evaluations

The different approaches to making LOPA consequence severity estimates follow the consequence screening methods described in Section 1.1 (Step 1):

- Estimate severity using consequences categorized in terms of the type and magnitude of a hazardous material or process energy release or other consequence characteristic. This approach has the advantages of simplicity and avoidance of needing to estimate impacts such as on people, but still requires some estimate be made of the release magnitude. However, it will not distinguish between the same release event being in a remote area at one facility and in a highly populated area at another facility. For this reason, the category distinctions may need to be site-specific. The magnitude of a release implies anticipated impacts; hence, the probability of realizing loss and harm is often implicitly included when setting up the release magnitude categories.
- Qualitatively estimate severity using impact categories in terms of human harm, environmental impact and/or business impact. The example in the preceding Table 1.1 illustrates this approach, where the LOPA team has assessed the vessel rupture consequence severity to be "Category 5" (which is perhaps a "High Severity" impact with potential for injuries and

fatalities). Such an evaluation might use conservative assumptions such as flammable releases always igniting and/or personnel always being present when a release, fire or explosion occurs.

- Estimate severity using qualitative or order-of-magnitude impact categories in terms of human harm, environmental impact and/or business impact, with adjustments for post-release probabilities. This approach includes an explicit, documented consideration of conditional modifiers such as the probability of ignition and the probability of personnel presence in the effect area.

- Quantitatively estimate human harm, environmental impact and/or business impact using detailed analyses in determining the characteristics of a release and its effects. Tools associated with full QRAs may be used here, including dispersion and blast effects analysis. Enabling conditions (where pertinent) and conditional modifiers are likely to be employed in this type of consequence analysis, as are more detailed estimates of the factors contributing to loss event frequency.

It needs to be emphasized that **the approach selected must be consistent with the risk criteria used in the LOPA for determining the adequacy of risk control measures.** This is further discussed in Section 1.4 in relation to risk criteria endpoints.

1.3   When to Use Enabling Conditions and Conditional Modifiers

Enabling conditions and conditional modifiers are not used in every LOPA. They only warrant being used when they support the objectives of the LOPA and are consistent with the risk criteria employed.

Guidance on When to Use Enabling Conditions

The following are typical situations where enabling conditions might be used in LOPAs:

- The event sequence would only be realized if the unit was in a particular state of operation (e.g., in recycle mode or feeding directly from a transport container), where the unit being in that state is necessary to realize a consequence of concern but independent of the rest of the event sequence.

- The consequence would only be realized if the unit was using a particular raw material or catalyst or processing a particular formulation, where the situation was necessary to achieve the consequence of concern but independent of the rest of the event sequence.

- The event sequence would only be realized if other circumstances such as a low or high ambient temperature existed at the time an initiating event occurred.

In each case, the purpose of employing the enabling condition is to take into account conditions that are necessary for an abnormal situation to proceed to the consequence of concern. The capabilities of the LOPA analyst(s), the established company or facility LOPA methodology and the availability of relevant data would all need to support the use of enabling conditions.

A full description of LOPA enabling conditions, along with worked examples, is given in Chapter 2. Enabling conditions may be associated with short-duration situations with severe potential consequences. The elevated risk exposure during these brief situations, termed "peak risk," should be considered and managed appropriately. This topic is discussed in Appendix B.

Guidance on When NOT to Use Enabling Conditions

The following are typical situations where a LOPA team should avoid the use of enabling conditions:

- The LOPA analyst(s) have insufficient knowledge of enabling conditions to employ them correctly.
- Insufficient data or information is available to assess the probability to be assigned to an enabling condition.
- The company's or facility's established LOPA procedure specifically indicates that enabling conditions are not to be used in its LOPAs, for whatever reason.
- A potential enabling condition does not meet company criteria. For example, established procedures may specify that an enabling condition is valid only if it gives a full order of magnitude reduction in frequency, but the potential enabling condition does not meet this.
- The company or facility does not have the resources, capability or practices in place to properly assess and document the use of enabling conditions and maintain their ongoing validity. (See Section 2.6 for guidance on maintaining the validity of enabling conditions over time.)

In addition, situations where one particular type of enabling condition, namely a "time-at-risk" enabling condition, should not be used are listed in Section 2.3.

Guidance on When to Use Conditional Modifiers

A typical situation where conditional modifiers might be used in LOPAs is when a company's risk criteria are based on best-estimate risk values rather than conservative bounding estimates. In this case, not using conditional modifiers may result in risk estimates inconsistent with the company's risk criteria. For example, in a situation where the probability of igniting a given release is expected to be much less than 100%, then using a probability of ignition of 100% would give a conservative risk estimate but not a best-estimate value. Basing risk management decisions on an overly conservative risk estimate could result in a misallocation of risk-reduction resources. For scenarios where more than one conditional modifier is pertinent, this over-conservatism could be even further amplified.

The capabilities of the LOPA analyst(s), the established company or facility LOPA methodology and the availability of relevant data would all need to support the use of conditional modifiers. A full description of LOPA conditional modifiers, along with worked examples, is given in Chapter 3.

Guidance on When NOT to Use Conditional Modifiers

Many organizations decide not to use conditional modifiers in LOPAs for various reasons that might include one or more of the following:

- If the organization's approach is used to implicitly include conditional modifier probabilities when selecting a consequence severity category, then their explicit use in a LOPA would be double-counting these factors.
- If the organization judges that the uncertainties or complexities involved in incorporating conditional modifier values in LOPAs are too great to warrant their use.
- If the organization considers that the difficulties inherent in validating conditional modifier values are considered to be too great, recognizing that conditional modifiers cannot generally be audited or functionally tested in the same way as for initiating events and independent protection layers.
- If the organization chooses to use a conservative approach that considers conditional modifier factors such as likelihood of ignition to always have a probability of 1.

In addition, the following are typical situations where conditional modifiers should be avoided or limited in LOPAs:

- The facility's risk criteria for evaluating LOPA scenarios uses severity categories based on the size of a material or energy release (for example, a 10,000 lb flammable liquid release may be one severity category) rather than the potential consequences of the release such as fire or vapor cloud explosion, fatalities or environmental impacts.
- The LOPA analyst(s) have insufficient knowledge of conditional modifiers to employ them correctly.
- Insufficient data or information is available to assess the probability to be assigned to a conditional modifier.
- The company's or facility's established LOPA procedure specifically indicates that conditional modifiers are not to be used in its LOPAs, for whatever reason.
- The company's or facility's established LOPA procedure is to not use conditional modifiers unless they provide a full order-of-magnitude effect on the risk calculation.
- The company or facility does not have the resources, capabilities or practices in place to properly document the use of conditional modifiers and maintain their ongoing validity. (See Section 3.8 for guidance on maintaining the validity of conditional modifiers over time.)

- Even when an organization's approach is to use conditional modifiers in LOPAs, situations exist where specific conditional modifiers should not be used when evaluating specific scenarios, as discussed in Chapter 3 for the various types of conditional modifiers.

It is hoped that the current *Guidelines* will aid in the decision process as to the appropriate use of enabling conditions and/or conditional modifiers in LOPAs and other risk evaluations, and to provide the understanding needed for their proper usage.

1.4 Risk Criteria Endpoints

As mentioned earlier, the consequence categories and risk criteria used in evaluating the adequacy of risk control measures must match the methodology used for estimating scenario risk. The basic difference between the categories and risk criteria used is the selection of *endpoints* for the determination of consequences. These endpoints can range from release magnitude to injury/fatality, environmental damage and/or business impacts or impact categories. This section will further discuss and illustrate different possible endpoints for various types of loss events (fires, explosions, toxic releases).

The following example will be used to illustrate different endpoints that may be used in a LOPA study. The hypothetical scenario for this example will be loss of containment of a toxic and flammable liquefied gas called "flamitox". The possible general outcomes of the loss-of-containment event are illustrated in the event tree of Figure 1.1. Table 1.2 shows the range of possible endpoints for addressing this event, following the same four consequence screening methods described in the LOPA overview in Section 1.1.

Note that this discussion of consequence endpoints is applicable to conditional modifiers but not to enabling conditions. Even if a LOPA does not include conditional modifiers (i.e., uses endpoint type 1 or 2), it may include enabling conditions when evaluating the scenario frequency.

As can be seen in Table 1.2, conditional modifiers are normally applied to analyses when the risk criteria are based on ultimate consequences, often fatalities. The general principle is to develop endpoints in concert with the methodology and the risk criteria for those endpoints. Referring again to the example in Figure 1.1 and Table 1.2:

- A company that uses endpoint type 1 (release magnitude) might have a risk boundary of a specified frequency of having a 10 t flamitox release; i.e., if the likelihood is greater than that frequency, then risk reduction is required; if less than or equal to that frequency, safeguards are considered adequate and no further risk reduction is required.

- Another company, which uses endpoint type 2 (impact categories with no consideration of conditional modifiers), might have the same risk boundary but apply it to scenarios with any potential for severe injuries or fatalities. Companies using this approach need to recognize implicit assumptions that are made, such as probability of ignition of flammable vapor clouds assumed as 100%, and whether the risk boundary is based on worst-case impacts or on most likely impacts.
- Yet another company, which uses endpoint type 3 (impact categories with conditional modifiers considered), might have a different risk boundary for having an incident with an expected severe injury or fatality impact, since additional probabilities may be included in the scenario likelihood such as the probability of a person being present in the effect area at the time of the incident. Note that this same risk boundary might be used by companies performing quantitative risk analyses.

The determination of who might be impacted by a LOPA scenario goes beyond whether or not conditional modifiers are employed. The following paragraphs discuss various means of determining toxic release, fire and explosion impact boundaries.

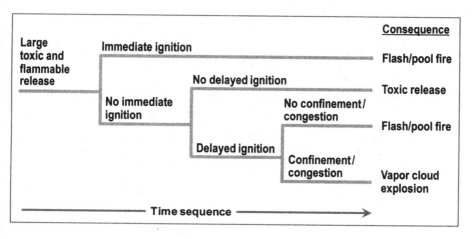

Figure 1.1 Event tree to illustrate different LOPA endpoints.

Table 1.2 Range of possible endpoints for example toxic/flammable release event

Endpoint type	Typical LOPA consequence description	Consideration of ignition probability	Consideration of other conditional modifiers	Number of scenarios evaluated
1 Release magnitude	10 t flamitox release	Not addressed	Not addressed	1
2 Qualitative categories of potential loss and harm impacts	10 t flamitox release, resulting in possible fire/explosion and potential for severe injury or fatality	Not explicitly considered; often assumed to be 100%	Reasonable worst-case consequences estimated, generally without explicitly considering, e.g., personnel presence or fatality probability	1
3 Qualitative or order-of-magnitude categories of potential loss and harm impacts, with conditional modifiers	10 t flamitox release generating a large toxic vapor cloud; severe injury/fatality from inhalation of toxic vapors	Probability of not igniting may be included in analysis	Conditional modifiers may be considered such as probability of fatality and either wind direction or personnel presence / time at risk	3
	10 t flamitox release resulting in flash fire and pool fire; severe injury/fatality from exposure to fire thermal radiation	Ignition probability may be included in analysis	Conditional modifiers may be considered such as fatality and personnel presence / time at risk probabilities	
	10 t flamitox release generating a large flammable vapor cloud migrating to confined/congested area(s); vapor cloud explosion with severe injury/fatality from exposure to thermal, blast effects	Probability of delayed ignition may be included in analysis	Conditional modifiers may be considered such as fatality and personnel presence / time at risk, as well as wind toward confined/congested area or probability of sufficient confinement and congestion	
4 Quantitative impacts, with conditional modifiers	Same as 3 above; may have delayed vs. immediate ignition scenarios for flash/pool fires	Ignition probability generally quantified	Conditional modifiers quantified, to the extent necessary to meet the analysis objectives	3 or 4

Toxic Release Impacts

Toxic release impact boundaries used in LOPAs range from exceeding a toxic threshold for sensitive individuals to exceeding the inhaled dose required to cause fatalities in a normal healthy worker population. The effect area and probability of exceeding the toxic threshold can be greatly different across this range of toxicity measures described in this section, so care should be taken that the risk criteria used by a company for toxic release effects are consistent with the toxic endpoint employed.

Duration of exposure is nearly as important of a consideration as vapor concentration. This is because the effect severity for most chemicals posing toxic inhalation hazards is based on the inhaled dose, which is a function of both concentration and time.

Toxic effect thresholds. Toxic release effects may be considered in a LOPA as resulting in human harm if a "serious health effect" concentration or dose threshold is exceeded. One set of toxic effect thresholds that can be used in determining impact boundaries where the general public could be affected are Acute Exposure Guideline Levels (AEGLs), established for over 100 substances at five exposure periods (10 and 30 minutes, 1 hour, 4 hours, and 8 hours). The second of the three AEGL levels, the AEGL-2, is defined as the airborne concentration of a substance above which it is predicted that the general population, including susceptible individuals, could experience irreversible or other serious, long-lasting adverse health effects or an impaired ability to escape.

Emergency Response Planning Guidelines (ERPGs) are defined similarly, but are based on a one-hour exposure duration. Counting any location at which the ERPG-2 is exceeded as being impacted by a given toxic release event is often quite conservative, especially with chemicals that have good warning properties. For example, the ERPG-2 for ammonia is 150 ppm, which is well above its odor threshold of 5 ppm. A person is very unlikely to be exposed to ammonia vapors for an entire hour without the vapors dissipating or exposed persons taking protective action. As a result, risk criteria based on the frequency at which one or more persons would be exposed to a vapor concentration above a toxic effect threshold such as the ERPG-2, regardless of the exposure duration, are likely to be different than risk criteria based on exceeding toxic effect thresholds for the actual expected duration of exposure.

Lethal thresholds. If a company's risk criteria are associated with human fatality potential, then a threshold lethal concentration or dose might be used. Typical thresholds include the AEGL-3, ERPG-3 and the LC_{LO} (lethal concentration threshold). The AEGL-3, for the same five exposure durations as the AEGL-2 (10 and 30 minutes, 1 hour, 4 hours, and 8 hours), is defined as the airborne concentration of a substance above which it is predicted that the general population, including susceptible individuals, could experience life-threatening health effects or death. By comparison, the ERPG-3 would be a conservative threshold for most scenarios, in that it is defined as the maximum airborne concentration for exposures up to an hour in duration without experiencing or developing life-threatening health effects. The LC_{LO} may likewise be conservative, in that it does not explicitly consider duration of exposure.

50% effect level. Rather than using a threshold concentration, if a company's risk criteria pertain to average or best-estimate health effects for a given population, then the LC_{50} (lethal concentration to 50% of a population for a given inhalation exposure time) might be used for the toxic release endpoint. The frequency of having a release large enough to exceed a 50% effect level would be generally lower than to exceed a threshold effect level. This again demonstrates the close connection between the LOPA method and risk criteria used.

Other toxic endpoints. Probit equations may be used in a quantitative risk analysis to calculate a probability of fatality as a function of inhaled concentration and duration for many volatile toxic materials. An approach that is intermediate between a concentration-based approach and a probit analysis has been developed and published by the U.K. Health and Safety Executive (HSE 2011). This approach calculates a Dangerous Toxic Load (DTL) value by multiplying the exposure duration times the vapor concentration to the nth power, with $n = 1$ for most materials but higher for some chemicals. The DTL is compared to either a Specified Level of Toxicity (SLOT) or a Significant Likelihood of Death (SLOD) threshold to determine expected effects to the general population, such as for providing land use planning advice. Values of n, SLOT and SLOD are given in HSE 2011 for over 200 toxics, along with explanatory information and information for the derivation of SLOT and SLOD values for chemicals which are not listed.

Escape from a toxic cloud. As described above, the effects of toxic inhalation exposures are based on the exposure time as well as the inhaled concentration. The use of AEGLs, a probit equation or the DTL may allow the user to establish an endpoint concentration for an expected exposure time, taking into account the ability of persons to escape from a toxic cloud or to otherwise limit their inhalation exposure time, based on their risk target severity (e.g., threshold or 50% level).

Because duration of toxic exposure is such an important factor in determining the outcome of a toxic release event, assumptions made regarding whether persons will escape from a toxic cloud will have a significant bearing on the appropriate risk boundary to use. Options include:

- Assume no credit for escape from a toxic cloud; persons in a given location would be exposed the entire time required for the toxic cloud to pass, unless perhaps they are only passing through a relatively small area such as on operator rounds or driving in a vehicle. This approach may be most appropriate for toxic vapors that do not have good warning properties, or that can readily disable a person (e.g., interfere with vision), or situations where the potentially affected population has no training or means of receiving communication of the situation or means of escape.

- Use an expected-average exposure duration based on an assumption that persons would take protective action if possible and escape from the toxic cloud. This assumption may be pertinent for toxic vapors that have good warning properties and where persons are not likely to be trapped in a location such as an elevated platform where escape is impaired.
- Take the probability of successfully escaping from a toxic cloud into account by including it in the LOPA as a conditional modifier. (Note that taking credit for a risk reduction factor of this nature would warrant an evaluation of all the steps involved in detecting, deciding and acting to successfully escape, as well as ensuring their ongoing validity such as through equipment integrity maintenance and personnel training and drills. This conditional modifier is further discussed in Section 3.6.)

Other considerations when evaluating escape from a toxic cloud and likely exposure duration include sheltering in place, vapor detectors, area alarms and the use of personal protective equipment such as escape respirators.

Indoor versus outdoor concentration. As a related consideration, when establishing toxic release endpoints and associated risk boundaries, a company may consider how to treat persons inside an occupied structure that could be affected by an external toxic vapor cloud. Again, multiple options are possible:

- Assume no credit for being indoors; persons inside the structure are assumed to be exposed to the same vapor concentration as the outdoor air. This approach may be most appropriate for toxic vapors that do not have good warning properties and for releases of relatively long duration, or that can disable persons quickly at low concentrations.
- Calculate the expected indoor concentration based on pertinent factors such as the building air exchange rate, then compare the calculated indoor concentration to the threshold or other toxic effect level used. This approach recognizes that the indoor concentration is often considerably less than outside the building, particularly for short-duration releases. Some dispersion modeling software packages such as ALOHA® (EPA 2012) include routines for evaluating indoor concentrations of toxic materials that are exposed to an external toxic cloud.
- Assume all persons inside the occupied structure are adequately protected against the toxic release and are not counted as being potentially affected by the release event. This assumption may be pertinent for short-duration releases and for protective designs such as a structure that has a system to ensure positive air pressure inside the building relative to the outside air, designed for the specific toxic hazard and included in the facility's asset integrity management program to ensure the system is properly tested and maintained on an on-going basis.

Fire and Explosion Endpoints

Much of the above discussion also pertains to fire and explosion endpoints. Exposure duration in particular is an important factor when determining burn injury severity. The effects of personal exposure to thermal radiation from fires or flares are well-established, with typical thresholds being second-degree or third-degree burn boundaries.   Assumptions range from no credit for protective action such as seeking shielding or shelter, to use of a conditional modifier for probability of shielding or escape, to assuming protective action would always be taken within a given amount of time such as 40 seconds. Persons inside buildings are generally considered to be shielded from external flash fire or pool fire effects. Protection is sometimes credited against flash fire events when proper flame-resistant clothing is worn (but note that even persons protected by flame-resistant clothing could be injured by inhaling hot product-of-combustion gases).

Endpoints for explosion events can be more complex, since multiple mechanisms for causing injury or fatality are possible, including blast wave direct impingement, missiles or flying debris, being thrown against a hard surface or sharp object, being knocked or startled off of an elevated work location, or being inside a building that is damaged or collapses. Thresholds used for explosion events are sometimes greatly simplified, such as using a 1 psi overpressure as an endpoint, with an implied threshold of serious injury potential for persons inside buildings not blast-resistant. (A somewhat higher threshold such as 3 psi may be warranted for persons not inside buildings.) Conditional modifiers that pertain to these kinds of explosion effects are discussed in Section 3.6.

2
LOPA Enabling Conditions

This chapter defines and illustrates *enabling conditions* as they may be used in Layer of Protection Analysis. It is not intended to include an exhaustive set of possible enabling conditions, but rather give sufficient information and examples that the user can clearly recognize and properly employ enabling conditions where they are warranted.

2.1 Definition and Defining Characteristics

An *enabling condition* is a condition that makes the beginning of a scenario possible. An enabling condition is neither a failure nor a protection layer. It consists of an operation or condition that does not directly cause the scenario, but that must be present or active in order for the scenario to proceed to a loss event. Note that mitigating factors, such as the probability of personnel presence or of emergency evacuation, are *conditional modifiers* (Chapter 3) and not enabling conditions.

The term *enabling event* is sometimes used for *enabling condition*. The term *enabling condition* is preferred, since enabling conditions are not generally events but rather conditional states.

2.2 Interrelationship with Initiating Event

An enabling condition is expressed as a probability. The combination of the enabling condition probability with the initiating event frequency must always be a frequency that represents the times per year an abnormal situation would be encountered that could lead to a loss event. Note that most LOPA scenarios will not have enabling conditions.

2.3 Time-At-Risk Enabling Conditions

One general type of enabling conditions involves the concept of *time at risk*. Time at risk is when an incident sequence may only be realized a certain fraction of the time when conditions are right for the event sequence to progress to a loss event. An underlying assumption for time-at-risk enabling conditions is that *only revealed failures can act as initiating events during time-at-risk conditions*. A *revealed failure* is one that may be immediately or almost immediately apparent through an alarm or indicator system. For example, a primary feed pump failing off during continuous operation of a process would be rapidly apparent by its effects on process parameters when the feed flow is lost.

By contrast, *unrevealed (latent) failures,* such as a bypass line plugging or freezing up or a shutoff valve failing stuck in the open position, could occur at any time and remain dormant while still being able to run the process. If an unrevealed failure occurred <u>before</u> the beginning of the time at risk, and was then made evident <u>when the time at risk began,</u> it could then serve as an initiating event for an incident scenario. In this case, time at risk should <u>not</u> be taken into account as a LOPA enabling condition. Time-at-risk considerations can only be applied as enabling conditions when systems have been put in place to reliably ensure that potential unrevealed failures that could lead to incident scenarios are detected and corrected before the beginning of the time-at-risk state, or when the failures are naturally revealed due to the design of the process.

Seasonal Risks

One common time-at-risk enabling condition is a sufficiently low ambient temperature to enable process or utility lines or instrumentation to freeze following failure of designed freeze protection. Enabling conditions may also involve other seasonal risks such as an extreme high ambient temperature affecting cooling capacity or a low-humidity condition allowing static electricity accumulation and discharge.

Table 2.1 shows an example of a LOPA scenario involving a time-at-risk enabling condition. The only LOPA worksheet portions shown are those involved in combining the initiating event frequency with the enabling condition probability to obtain the loss event frequency without conditional modifiers or independent protection layers.

Note that, in the example of Table 2.1, the 0.02 enabling condition probability would presumably have a documented basis or rationale, such as from actual climatological data. Note also that the approach used in this LOPA is to not restrict values to order-of-magnitude numbers, as discussed in Section 1.2. Climatological data is likely to support a more refined estimate of this particular factor than an order-of-magnitude probability. (The overall uncertainty in the scenario frequency will probably be determined by the uncertainty in the initiating event frequency rather than in the enabling condition probability.)

Table 2.1 Time-at-risk enabling condition example

		Frequency	Probability
Scenario title: Freeze-up of bromine transfer line from storage resulting in interruption of bromine feed to bromination reactor, unreacted organic feed passing to downstream process, ignition of organic vapors in final product storage tank head space and tank rupture with release of contents and impacts on personnel and the environment.			
...			
Initiating event	Plant-wide loss of steam supply, including the steam supply to the steam tracing on the bromine transfer line from storage	0.1/yr	
Enabling condition	Fraction of time ambient temperature is below -7 °C freezing point of bromine at plant location		0.02
Conditional modifiers	None		1
Frequency of unmitigated consequence		0.002/yr	
...			

Caution must be exercised when evaluating seasonal risk scenarios due to the possibility of <u>common mode failures</u> between the enabling condition and an independent protection layer. True independence is needed for a time-at-risk enabling condition. The following examples illustrate possible common-mode situations:

- A low-temperature condition both allowing an abnormal situation to develop and causing failure of a protective system due to instrumentation freeze-up.
- A rainstorm both enabling an operational error to occur due to poor visibility and contributing to an ensuing release by secondary containment overflow.
- Low ambient temperature causing a higher steam demand (for building and process heating) while also causing more stress and a higher failure rate for steam system components, if the high steam demand mode is the time at risk under consideration for an enabling condition.

Also, the seasonal condition should not by itself be the scenario initiating event, if it is to be considered an enabling condition. For example, if a facility on a waterway is designed for a hundred-year flood but a weather condition results in a 500-year flood that floats a tank and causes a release, then the extreme weather condition is the initiating event. Likewise, a storm producing winds exceeding the facility's design wind speed and resulting in a release or other process or control failure would be an initiating event and not an enabling condition.

Process State Risks

Another type of time-at-risk enabling condition is when a process must be in a certain part of a non-continuous operation when a failure occurs for the incident sequence to be able to proceed to a loss event. Three examples are given as follows.

Example 1. A batch chemical reaction may have potential for a runaway reaction, but only if loss of cooling occurs during the first part of the batch when most of the conversion takes place. If loss of cooling occurs during any other part of the batch sequence, no runaway reaction would result. In this example, the fraction of time the reactive system would be in a state such that loss of cooling could lead to a runaway reaction could be used as an enabling condition probability. Note that the loss of cooling cause and its likelihood of occurrence must be independent of the process state for the enabling condition to apply. In addition, the loss of cooling must be a *revealed failure* prior to the reaction entering the first part of the batch. Otherwise, the loss of cooling could occur at any point in the batch but would only be revealed when the reactor enters the first part of the batch and experiences a runaway reaction.

Example 2. If a particular isolation valve fails to close during a batch sequence, it could result in backflow through the valve leading to a consequence of concern, but only if the connected process is in a state where it is pressurized when the valve failure occurs. If the connected process is pressurized only a fraction of the time and this pressurization is independent of the state of the process upstream of the isolation valve and independent of the batch sequence, then this fraction of time could be considered as an enabling condition.

Example 3. The peak pressure that can be generated inside an enclosed vessel upon ignition of a flammable vapor/air mixture inside the vessel depends on the initial pressure at the time of ignition (as well as other factors such as how close the vapor is to a stoichiometric mixture). For instance, assume the vessel strength is such that it can withstand the peak pressure of an internal deflagration under normal conditions but not if the internal pressure was elevated at the time of ignition. If the vessel is used for a batch operation that has a high-pressure step that is only a fraction of the entire batch sequence time, then the fraction of time the operation is in the high-pressure part of the batch could be used as an enabling condition if the time at which the vapors ignite is independent of whether or not the batch is in the high-pressure part of the batch sequence.

Scenarios must be evaluated carefully to determine whether an enabling condition is involved. For example, for a chlorine railcar unloading operation, one scenario might be rupture of an unloading hose during chlorine transfer due to mechanical failure (i.e., not due to an external force such as railcar movement, and not due to a human error such as blocking in the liquefied gas). For example, if transfers are actually in progress 16 hours per week, then it may be tempting to multiply the failure rate of the hose by a process-state enabling condition probability of:

$$16 \text{ transfer-hours per week} / (168 \text{ h/wk}) = 0.1$$

However, failure of the unloading hose is more likely to occur during transfer when the system is pressured up than when the hose is not in use. In this situation, it would only be appropriate to use the enabling condition probability *if the hose was pressure-tested prior to each usage*.

Caution regarding failure rate data employed. In addition, it is important to determine the conditions under which failure rate data were taken. If, in the above example, the hose failure rate data were obtained from hoses that were typically in service 16 transfer-hours per week, then no enabling condition probability should be used, even if pressure-tested prior to each usage. On the other hand, if the hose failure rate data were obtained from hoses that were in service essentially 168 hours per week, then the enabling condition probability may be appropriate. If the hose failure rate data were based on number of pressurization cycles, then the number of pressurization cycles per year in the actual facility should be compared to the number of cycles per year for the failure rate data. The discussion in this paragraph is another example where incorrect use of an enabling condition probability can result in underestimation of the scenario frequency.

Situations Where Time At Risk Should NOT Be Used For Enabling Conditions

The following are situations where use of a time-at-risk enabling condition in a LOPA is generally not appropriate:

- Continuous operations where hazards are constant over time.
- *Infrequent, short-duration operating modes with high potential severity.* Use of a time-at-risk enabling condition in such a situation may lead to a conclusion that the average risk meets a facility's risk criteria; however, the peak risk imposed on persons exposed to the hazard may be inappropriately high. See Appendix B for a discussion of peak risk concepts.

  Example. Without nitrogen purging, the composition of vapors in the vapor space of a flammable liquid storage tank would pass through the flammable range each time the tank is put into service or taken out of service. Even though the fraction of time the tank vapors are in the flammable range might be very low (for example, 1 hour per year), personnel are likely to be in the vicinity during cold start and shutdown phases, and the peak risk imposed on operating personnel during that time period could be quite high.

- Procedure-based operations (sometimes called "transient operations") involving human error in the execution of a stepwise procedure. In such scenarios, the initiating event frequency is generally a combination of the frequency of conducting the procedural step in question with the probability of the specific human error when performing the step. Because a failure probability is involved, this does not meet the definition of an enabling condition. Recall that part of the definition of an enabling condition is that it is not a failure.

  Example. Consider a compressor overspeed trip test scenario where improper execution of the overspeed trip test could lead to equipment damage and possible personnel injury. The initiating event frequency in this case is a combination of the frequency at which the overspeed trip test is conducted (e.g., one test per year) and the probability of improper execution of the test (e.g., probability of 0.01 per test). No enabling condition is involved.

CAUTION: If taking enabling condition credit for an at-risk time fraction such as a process being in a certain step or making a certain product, the analyst should ask the following question. If the initiating condition occurs when the process is NOT at risk, will the failure be detected before the process next enters the at-risk phase? If the answer is *yes*, the failure will be detected and an enabling condition probability may be appropriate. If the answer is *no*, then it may be erroneous to consider this an enabling condition, since the failure would only be detected when the consequence of concern occurs. (In this case, it may be an initiating event and not an enabling condition.)

- Continuous or procedure-based operations involving any combination of factors that are failure or fault conditions, regardless of whether those conditions are revealed or latent. Whether a factor is an enabling condition or not may be determined by the actual scenario. The following examples, involving tank overfill scenarios as shown in Figure 2.1, illustrate this point.

  Example 1. If the scenario being evaluated is tank overfill due to a fixed quantity of liquid such as the contents of a 55 gal (208 L) drum being added to the wrong 200 gal (757 L) tank, then if the tank being overfilled is always kept full, there is no enabling condition, and the unmitigated overfill frequency (i.e., not taking any independent protection layers into account such as overfill protection) is calculated as follows:

  Tank overfill frequency = Tank selection error frequency

Example 2. If the same scenario is evaluated, except the wrong tank can have any amount of liquid in it, from empty to full, then an enabling condition can be used. The enabling condition probability is the fraction of time the tank level is high enough that the tank will be overfilled when the fixed quantity of liquid is added to it (taken as $55/200 \approx 0.3$). If the starting level is random and independent of the initiating event, then the unmitigated overfill frequency is:

Tank overfill frequency = 0.3 · Tank selection error frequency

Note that the enabling condition probability of 0.3 does not provide a full order of magnitude of risk reduction, so it might not be used in LOPAs employing strict order-of-magnitude approaches.

- Incident sequences where time at risk applies to the severity of consequences of the loss event, rather than its likelihood of occurrence. In such cases, the probability factor should be considered a conditional modifier, as discussed in Chapter 3.

    Example 1. Consider a situation where only 1% of the time personnel are close enough to a hazardous process to be affected by a loss event, such as a liquid leak into a confined area. Although this is a time-at-risk probability, it is a conditional modifier and not an enabling condition.

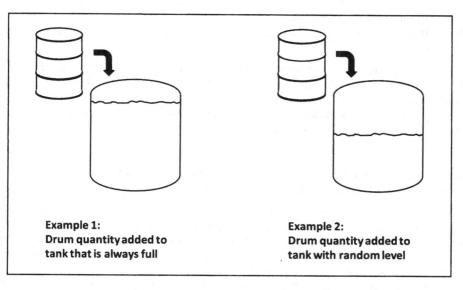

Figure 2.1 Diagrams for time-at-risk enabling condition examples.

Example 2. Consider a scenario involving a non-aerosolized overfill of an uninsulated *n*-octane storage tank into a concrete diked area. With a flash point of 13 °C (56 °F), the scenario has the potential for the spill igniting and resulting in a pool fire only during the fraction of the year that the ambient temperature is above the *n*-octane flash point. This situation involves a probability-of-ignition conditional modifier and not an enabling condition.

In general, if a time-based probability is associated with the part of an incident sequence leading up to a release of hazardous material or energy, it is in the realm of enabling conditions, whereas if the time-based probability is associated with the post-release part of the incident sequence, it is in the realm of conditional modifiers.

2.4 Campaign Enabling Conditions

Similar to time-at-risk enabling conditions are *campaign* enabling conditions. Campaign enabling conditions are associated with processes that may differ from time to time or from batch to batch with respect to raw materials (chemicals, concentrations, rates, quantities), catalysts, final products, operating conditions and/or process configuration (e.g., recycle vs. non-recycle mode of operation), and these differences result in non-uniform risks during the different campaigns. The use of enabling conditions is one means of addressing the non-uniform risks in such facilities. The situation shown in Table 2.2 is an example of a campaign enabling condition.

Several types of campaign enabling conditions are described in the following paragraphs. Campaign enabling conditions should be evaluated and applied with caution. These enabling condition probabilities can change over time depending on factors such as market conditions, capacity utilization and the economics of operating in different campaign modes.

Equipment Failures That Would Only Lead to a Loss Event If the Process Is In a Particular State or Configuration

The fraction of time the process is in the given state or condition can be considered an enabling condition, *if* the amount of time the process is in that state or condition is more than a brief or transient situation, and the equipment failure is independent of the process being in that state or condition. Note that enabling conditions that may be associated with startup and shutdown modes of operation are generally considered "time-at-risk" rather than campaign enabling conditions, as described earlier in this chapter.

Table 2.2 Campaign enabling condition example

		Frequency	Probability
Scenario title: Cooling water failure results in runaway reaction with potential for reactor overpressure, leakage, rupture, injuries and fatalities. Agitation assumed. Also assumed: Loss of cooling water can be detected before the reactor begins the condition where runaway reaction is possible.			
...			
Initiating event	Loss of cooling water (revealed failure)	0.1/yr	
Enabling condition	Probability that reactor is in a condition where runaway reaction can occur on loss of cooling (average of about two one-week campaigns per month on an annual basis)		0.5
Conditional modifiers	None		1
Frequency of unmitigated consequence		0.05/yr	
...			

As an example, consider a control loop failure that could result in a loss event upon failure of independent protection layers, but only if the control loop failure occurred during a certain mode of operation such as when product is being recycled instead of being fed forward. The fraction of time the facility is in the recycle mode of operation can be considered an enabling condition. *Caution*: Do not use time at risk if the control loop can fail and remain uncorrected before the pertinent mode of operation is initiated.

The fraction of time used for determining this type of enabling condition probability is usually calculated on an annual-average basis. For example, if the operation is only in this recycle mode once a year for ten hours, then the enabling condition (to the nearest order of magnitude) is only

(10 h in recycle mode/yr of operation)/(8766 h/yr) = 0.001

The likelihood of the control loop failure occurring within this brief time window, assuming the failure is independent of the operating mode, is obviously very small and could usually be neglected. (However, if the consequences are very severe, then the combination may still need to be evaluated. The techniques described in Appendix A can be employed, and the concept of "peak risk" as described in Appendix B may need to be considered.)

By contrast, if in the above example the facility is in recycle mode ten times a year for an average of 100 hours each time, the enabling condition probability becomes much more significant:

(100 h/recycle)(10 times in recycle mode/yr)/(8766 h/yr) = 0.1

Note that some companies might limit the amount of risk-reduction credit allowed for an enabling condition of this nature, such as not allowing a campaign enabling condition probability less than 0.1.

CAUTIONS
- *Using an enabling condition that results in multiple orders of magnitude risk reduction must be done with great care to ensure it is a genuine factor that will apply on an ongoing basis.*
- *An enabling condition probability resulting in multiple orders of magnitude risk reduction may be a genuine factor but may impose a brief but high risk to an operations or maintenance staff person that should be reviewed to ensure the peak risk during the brief campaign time window is not excessive (see Appendix B).*

Another enabling condition example of a process needing to be in a particular state of operation for an incident sequence to proceed to a loss event is the fraction of time an off-gas stream might be directed to the atmosphere instead of being directed to a flare or thermal oxidizer, such that if an abnormal situation is initiated that results in the generation and venting of flammable gases, a release of flammable gases to the atmosphere would result unless a preventive safeguard intervenes (see Figure 2.2). This example would only apply in a situation where atmospheric venting of the normal off-gases was allowed (such as for temporarily shutting down the thermal oxidizer for routine maintenance), and the relief system as a whole was properly designed for the specific scenario being evaluated. The enabling condition in this case would be the fraction of time gases are diverted directly to the atmosphere instead of to the thermal oxidizer. (Note that this enabling condition would not apply if the initiating event could cause the enabling condition to occur. In this example, if the high vent flow to the thermal oxidizer caused by the abnormal situation could result in the thermal oxidizer shutting down and forcing the feed to the thermal oxidizer to vent, the enabling condition would not be independent of the initiating event and should not be counted.)

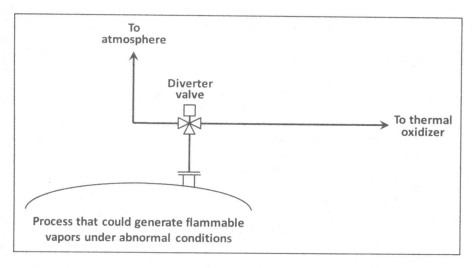

Figure 2.2 Diagram for campaign enabling condition example.

Facilities Operated Only Part of the Year

Similar to the above situation is a consideration of a facility or part of a facility that, for whatever reason, is not in operation a significant portion of the year (e.g., at least half of the year). If the failure rates employed for analyzing initiating event frequencies are still appropriate for a facility that is only in operation part of the year, then the fraction of time the facility is in operation can be considered an enabling condition probability. Table 2.3 exemplifies this type of enabling condition.

Table 2.3 Partial-year-operation enabling condition example

Scenario title: Loss of vacuum during operation of solvent purification facility results in sufficient solvent vapors being present to result in internal vapor/air deflagration and containment rupture if loss-of-vacuum safety system fails to protect and ignition source is present.			
		Frequency	**Probability**
...			
Initiating event	Vacuum pump fails off	0.1/yr	
Enabling condition	Solvent purification facility is in operation only 35 to 40 days per year; 40/365 days/year ≈ 0.1		0.1
Conditional modifiers	Ignition source present – *See Chapter 3*		1
Frequency of unmitigated consequence		0.01/yr	
...			

The following notes and cautions apply here:

- This enabling condition does not apply to scenarios involving human error during the execution of a procedure such as bringing the process back on line or mothballing it.
- The system must be in operable condition when restarting. Operation of a facility only part of the year should not be considered an enabling condition if the initiating failure for the scenario being studied can occur at any time of the year <u>and</u> remain undiscovered when the facility is restarted (i.e., an *unrevealed* or *latent failure*).
- Personnel are assumed to have received refresher training as needed before restarting.
- If the fraction of time the facility is in operation changes significantly, the scenario risk will be affected. Hence, this risk impact would need to be evaluated as part of the potential impact on safety and health when the operational change is examined as part of the facility's management of change process.

Chemical Processes Involving a Family of Reactive Chemistry or Other Composition-Related or Process-Related Variables

Specialty chemical producers often manufacture a range of products depending on customer demand and availability of raw materials. Refineries and hydrocarbon processing facilities may vary a product mix depending on such factors as incoming crude composition, market conditions, or time of year. Pharmaceutical producers may need to go through more hazardous reaction steps or use more hazardous raw materials for some products than for others. For a given family of raw materials, reactive chemistry and/or similar products, some applications may pose a particular hazard while others in the same family may have greatly reduced hazards depending on such molecular factors as carbon chain length, straight-chain vs. branching, oxygen balance or degree of halogenation; or such hazard characteristics as volatility, concentration, decomposition temperature or benzene content; or processing necessities such as operating above the boiling point or autoignition temperature of a process material.

When the fraction of time a batch chemical operation or other process facility uses hazardous materials or operates under conditions such that a given loss event is possible, and the loss event potential is nonexistent or greatly reduced the remainder of the time, then an enabling condition probability can be used in a LOPA to represent the fraction of time the loss event is possible, such as was illustrated in the previously presented Table 2.2.

Some cautions are warranted when employing an enabling condition in these situations:

- Since the enabling condition represents the probability that a more severe consequence (greater impact) could be realized, ensure the consequence severity for the LOPA scenario is consistent with the more severe consequence, not an average over the range of possible consequences.

- If the loss event impact is <u>not</u> significantly reduced (e.g., by an order of magnitude or more) the remainder of the time, then the use of an enabling condition may not be appropriate.
- It is likely that situations such as product variations, customer demands, market conditions and raw material availability will vary over time. If these have a significant effect on the product mix, then an evaluation of the risk impact of changing the product mix should be captured as part of the facility's management of change.

2.5 Other Possible Enabling Conditions

Other enabling conditions are possible that are not specifically time-at-risk or campaign situations as discussed above, but still involve a probability of a certain non-failure condition existing for the incident sequence to proceed to a loss event. An example is shown in Table 2.4 where a certain class of chemicals must be in storage for an initiating event in a refrigerated storage area to proceed to a reactive chemical incident, but that class of chemicals is only stored there 30% of the time. This example assumes the refrigeration system is supposed to be in operation at all times. (Note that this example could possibly be considered either a time-at-risk or a campaign enabling condition. Note also that the fraction of time temperature-sensitive materials are stored should be subject to management of change.)

2.6 Documenting and Validating Enabling Conditions

The examples shown in Tables 2.1 through 2.4 illustrate only one way of documenting enabling conditions in LOPAs. Some companies may require more than the enabling condition description and probability in the LOPA documentation, such as source references or calculations to back up enabling condition probabilities. The same is true if a range of possible values are associated with a given LOPA factor.

Table 2.4 Other enabling condition example

Scenario title: Thermal decomposition of temperature-sensitive organic peroxides in storage leads to container ruptures and a large fire/explosion event.		Frequency	Probability
...			
Initiating event	Cold storage unit refrigeration failure	0.1/yr	
Enabling condition	Historically, ~ 110 days per year, organic peroxides are stored in cold storage unit that require refrigeration to avoid thermal decomposition		0.3
Conditional modifiers	None		1
Frequency of unmitigated consequence		0.03/yr	
...			

For example, in the organic peroxides example of Table 2.4, supporting documentation could show the range of organic peroxide formulations considered to be part of normal storage operations, the self-accelerating decomposition temperature (SADT) or maximum safe storage temperature (MSST) of each formulation, and inventory control summaries showing storage frequencies and durations for each formulation. In any case, the enabling condition description in the LOPA worksheet can include a note to indicate whether the probability is an estimated value or is based on actual data.

In addition to the documentation of enabling conditions and their respective probabilities in a LOPA, an organization may incorporate a further step of validation to its basic LOPA approach to ensure that the selected enabling condition values are appropriate. For example, if a LOPA is performed at the final design stage of a new project, assumptions might need to be made regarding the expected fraction of time the process will be in a particular state where an incident sequence could proceed to a loss event. After the facility is constructed and commissioned, this fraction of time can be validated by comparison with actual operating experience and the LOPA risk calculations adjusted if necessary to reflect the actual experience. This may need to be integrated with the facility's management of change program to capture any potential safety and health impacts of deviating from expected operation.

An organization's approach to documenting and validating LOPA enabling conditions can be incorporated into its procedure (rule set) for how LOPAs are to be conducted within the organization. A partial example of such requirements is presented in Appendix C.

3

LOPA Conditional Modifiers

This chapter defines and illustrates *conditional modifiers* as they may be used in Layer of Protection Analysis. It is not intended to include an exhaustive set of possible conditional modifiers, but rather give sufficient information and examples that the user can clearly recognize and properly employ conditional modifiers where they are warranted. Following a general discussion of conditional modifier characteristics, the sections in this chapter cover five specific types of conditional modifiers.

3.1 Definition and Defining Characteristics

A conditional modifier is one of several possible probabilities included in scenario risk calculations when risk criteria endpoints are expressed in impact terms (e.g., fatalities) instead of in primary loss event terms (e.g., release, vessel rupture). Conditional modifiers include, but are not necessarily limited to:

- Probability of a hazardous atmosphere
- Probability of ignition or initiation
- Probability of explosion
- Probability of personnel presence
- Probability of injury or fatality
- Probability of equipment damage or other financial impact

"Probability of environmental impact" would also be a possible conditional modifier. If used, it could follow the same general approaches as probability of injury or fatality and/or probability of equipment damage or other financial impact.

Conditional Modifiers vs. Independent Protection Layers

Both conditional modifiers and independent protection layers (IPLs) are risk reduction factors, expressed as probabilities, which apply to specific scenarios. However, IPLs are engineered and/or administrative safeguards that are capable of responding to a process deviation to either avoid or mitigate a loss event. Examples include effective operator response to an alarm, safety instrumented functions, emergency relief systems, and source-mitigative engineered safeguards such as excess flow valves and detection/isolation systems.

Conditional modifiers, by contrast, are not response actions but relate to the condition that the facility is likely to be in at a particular point or period of time in an incident sequence. The conditional modifier probability is the fraction of time the facility is in that condition when the incident sequence arrives at the pertinent step in the sequence. For example, if a loss event occurs generating a large flammable vapor cloud, then at that point or period of time in the incident sequence either a sufficiently energetic ignition source will be present within the flammable envelope of the cloud before the cloud dissipates, or no sufficiently energetic ignition source will be present. The conditional modifier probability expresses the fraction of time (or, at least, an estimate of that time fraction) that such an ignition source would be present.

Note that even though conditional modifiers are not response actions, they can nevertheless be affected by how a facility is managed and maintained. For example, the probability of personnel being present when a material or energy release occurs can be affected by such factors as hazardous area access control and routing of vehicle and pedestrian traffic. The probability of ignition within an electrically classified area can be significantly increased by not inspecting and maintaining purged enclosures and sealed conduits or by allowing general-purpose equipment to be in the area.

Characteristics of Conditional Modifiers

Some of the same characteristics apply to conditional modifiers as do to IPLs; namely, independence, auditability and the need to identify and manage changes that could affect conditional modifier probabilities. Examples of the need for independence are included in the following paragraphs and examples. Auditability and the documentation of conditional modifiers are discussed in Section 3.9.

Independent of the severity estimate. When conditional modifiers are used, they are considered as probabilities that are independent of other considerations and factors in the scenario risk calculations. In this regard, particular care should be taken when assessing the severity of loss event consequences, as illustrated in Example 3-1.

Example 3-1 Independent of the severity estimate

Scenario	A dust collector on the outside of a process building is close to a walkway between the contractor parking lot and entry gate. If a dust explosion ruptures the collector enclosure and connecting ductwork, recordable injuries are likely inside the building and severe or fatal injuries are likely during the approximately 45 minutes per day that pedestrians are using the walkway near the dust collector.
Correct conditional modifier	For the more significant consequence of severe or fatal injuries to occur, personnel must be present outdoors near the dust collector at the time of the explosion. Since this is true about 45 minutes (0.75 hour) out of each 24 hours, the conditional modifier probability for personnel presence is 0.75 h / 24 h = 0.03 and the impact is severe or fatal worker injuries.
Wrong way example	If the conditional modifier of 0.03 as calculated above is employed, then a lesser impact of recordable injuries or a weighted-average impact between recordable and severe/fatal injuries should not be used, since this would be double-counting the factor that most of the time no pedestrians would be impacted.
Alternative approach	The above situation could be broken down into two scenarios. One scenario would have the most-likely impact of recordable injuries with a probability of personnel presence of 1 (since persons are assumed to be always present inside the building), and the second scenario would have a severe/fatal injury impact and a 0.03 conditional modifier for probability of personnel presence as calculated above.

Independent of all other factors in the scenario. A conditional modifier is a probabilistic condition that is not intertwined with the scenario itself. In Example 3-1 above, it is reasonable to expect that there is no association between the contractors being on the pedestrian walkway outside the building and the scenario leading up to rupture of the dust collector enclosure. Example 3-2 describes an incident where the presence of personnel was intertwined with the incident scenario and not just a probabilistic condition.

Example 3-2 Independent of all other factors in the scenario

Scenario	In a 2008 incident at a facility in Institute, West Virginia, an uncontrolled decomposition of methomyl (a carbamate insecticide) resulted in a residue treater vessel rupturing. The explosion and ensuing fire resulted in two outside operators being fatally injured.
No conditional modifier	Ten minutes before the vessel ruptured, one outside operator was asked to go out and check the residue treater vent and the second operator was asked to assist in troubleshooting an abnormal process situation. Since the operators being in the area was associated with the actual scenario, the presence of personnel in the area at the time of the vessel rupture explosion is not a condition of the facility independent of the incident scenario, so a conditional modifier should not be employed.

In general, any detected equipment failure may require troubleshooting, so any scenario involving a detectable equipment failure should assume a 100% probability that personnel will be present (unless there is no time delay at all between the equipment failure and the loss event). Also, the occurrence of some events such as small leaks or fires can increase the probability of personnel being within the effect area of a loss event. These issues are further discussed in Section 3.5, Probability of Personnel Presence.

Use of Conditional Modifiers and Associated Pitfalls

If conditional modifiers are to be employed in LOPAs, their use should be intentional, disciplined and consistent. A good approach is for a company or facility to have a standardized baseline set of conditional modifiers, along with rules for their use. Such an approach can help avoid some common pitfalls in the usage of conditional modifiers, including:

- Using more conditional modifiers than typical incident scenarios truly warrant
- Being overly optimistic in estimating conditional modifier probabilities
- Including more conditional modifiers until a desired risk level is reached
- Using conditional modifiers to justify avoiding a standard practice such as not including overflow protection in a tank design
- Using conditional modifiers in situations where the facility's risk criteria assumed worst-case impacts with no conditional modifiers.

Conditional modifier probabilities can change significantly over time. For example, relocation of buildings or walkways can affect personnel presence and probability of injury or fatality. Rerouting traffic or adding outdoor smoking areas can change ignition probabilities. Changes in capacity utilization can have various effects, such as on staffing levels. Slight changes in reactive mixture formulations can alter the energy input required to initiate an uncontrolled reaction.

Thus, systems need to be in place to manage these changes and their effects on process risk.

Deterioration of assumed conditions also needs to be avoided. For example, a prerequisite to using a low probability of ignition for a flammable release due to the area having electrically classified equipment would be that the electrical equipment is maintained over time to avoid degradation of the equipment such as junction boxes, sealants, purges, lighting fixtures. Also, other ignition sources, such as smoking and general-duty electrical equipment, would need to continue to be excluded from the area for the conditional probability to remain valid.

CAUTION:

Make sure that if the LOPA team selects a conditional modifier probability, it is realistic to expect that the probability will be valid over time.

The use of conditional modifiers can expand the scope of a company's management of change (MOC) program. It may need to include aspects of the operation not normally considered to be process changes, such as those described in the preceding paragraphs.

As an example, a capital project constructing a new unit at an existing facility could require that an open plot next to a hazardous process be used as a laydown yard during construction. If a LOPA for the hazardous process considered probability of personnel presence, the additional occupancy caused by the adjacent plot being used as a laydown yard could result in the risks of the operating facility not being adequately controlled during construction of the new unit. The MOC system for the facility would need to be robust enough to capture this change and put in place any required compensating measures to address the increased risk.

Another example would be if housekeeping activities detected cigarette butts within a possible effect area for a flammable release scenario. Until this situation was positively corrected, including any underlying human factors issues, a near-miss incident investigation and/or a MOC review should look at any pertinent LOPAs in light of whether too much risk-reduction credit might have been given for probability of ignition or probability of personnel presence and whether any interim (or permanent) risk-reduction measures may be required.

Several of the most common types of conditional modifiers are discussed in the following sections.

3.2 Probability of a Hazardous Atmosphere

This term is used for conditional modifiers involving loss of primary containment (LOPC) events or operational upsets that may or may not result in a hazardous atmosphere being formed, depending on the actual conditions. "Hazardous atmosphere" can refer to a toxic, oxygen-deficient or oxygen-enriched atmosphere to which personnel could be exposed, or to a flammable vapor or explosible dust atmosphere.

For example, consider a small process building containing analytical instrumentation that relies on nitrogen for normal operation. This structure is not continuously occupied; however, it is entered several times per shift by operating personnel to record process measurements.

An asphyxiation hazard exists due to the use of nitrogen inside the building. The enclosure is equipped with a continuously operated ventilation system designed for temperature control purposes only. Nitrogen is supplied to the analytical instruments via small-bore tubing, with the primary source of nitrogen being external to the building. If a nitrogen line or connection fails inside the building, an oxygen-deficient atmosphere may or may not be created, depending on such factors as the type of component that has failed, the hole size associated with the LOPC event, and the operating pressure of the nitrogen. If this scenario was evaluated using LOPA and used "Nitrogen primary containment failure inside the enclosure" as the initiating event, then the initiating event frequency should be a total frequency of all nitrogen LOPC events inside the enclosure. However, as previously stated, not all such nitrogen leaks will create an oxygen-deficient atmosphere, so a "probability of hazardous atmosphere" conditional modifier in the LOPA could be used to represent a best estimate of the fraction of leaks that would be expected to create a hazardous atmosphere. Without this conditional modifier, every nitrogen leak included in the initiating event frequency would be assumed to create an oxygen-deficient atmosphere. (For a constant-pressure nitrogen source, an alternative would be to determine a minimum leak size that could create an oxygen-deficient atmosphere, then only include in the initiating event frequency those LOPC events greater than or equal to this minimum leak size.)

LOPC events involving high-flash-point flammable liquids are another candidate for use of a "probability of hazardous atmosphere" conditional modifier. Consider the example of Table 3.1 involving mixed isomers of xylene (i.e., "xylenes"), a flammable liquid with an open-cup flash point around 26 °C (78 °F). Scenarios involving flammable liquid LOPC leading to the formation of an outdoors liquid pool would most likely consider the consequences of a pool fire. However, in order to have a pool fire consequence, the xylenes must be above their flash point temperature to generate an ignitable vapor/air mixture (i.e., "hazardous atmosphere") above the liquid surface. (This is not always the case; for example, an ignitable fine mist could be formed by a pressurized release through a small orifice, or a hot surface could be present to heat up the xylenes, or a material may self-react or react with area moisture upon being released to the environment.) For an ambient-temperature spill, whether or not flammable vapors will be generated is thus a function of the ambient temperature at the time of the release. The use of a conditional modifier for probability of hazardous atmosphere, based on the fraction of time the ambient temperature is above the xylenes flash point, would provide a more accurate risk estimate than conservatively assuming all spills would be ignitable, particularly in colder climates. (Note that this could also be considered part of a probability-of-ignition conditional modifier.)

Table 3.1 Probability of hazardous atmosphere conditional modifier example

Scenario title: Release of mixed xylenes from blending unit, leading to formation and subsequent ignition of a liquid pool within the process unit, resulting in minor to moderate equipment damage as a result of the pool fire.		Frequency	Probability
...			
Initiating event	Loss of primary containment due to primary transfer pump P-101 mechanical seal failure	0.1/yr	
Enabling condition	None		1
Conditional modifiers	Probability of formation of a flammable atmosphere: ~ 6 months out of the year, ambient temperatures are greater than the open-cup flash point of xylenes; no mist formed		0.5
	Probability of ignition: Xylenes liquid release in a Class I, Division 2 area, based on 10 years of historical experience at this location (two of five xylene spills ignited)		0.4
Frequency of unmitigated consequence		0.02/yr	
...			

A third and somewhat unique example involves conveying combustible plastic resin pellets that have been impregnated with a blowing agent (e.g., expandable polystyrene, or EPS). These plastic pellets often contain a relatively low proportion of a flammable substance, such as pentane or isopentane, as the blowing agent. Following the polymerization and impregnation stages, it is common practice to convey these pellets through enclosed ductwork with either air or nitrogen as the motive fluid. In the event of a mechanical failure that results in loss of flow, the stationary pellets will very slowly evolve the flammable blowing agent. The evolution rate of the blowing agent diffusing from the pellet matrix is generally slow enough that several hours may elapse before a flammable atmosphere can be created inside the ductwork (assuming adequate mixing and a homogeneous atmosphere). In these scenarios, a conditional modifier can be used in the LOPA that takes into consideration the conditions necessary to create the flammable atmosphere.

Other situations involving probability of hazardous atmosphere are also possible, such as sufficient dust accumulation inside an enclosure to form an explosible dust cloud if dispersed within the enclosure. When employing a probability of hazardous atmosphere, it should be ensured that conditional factors are not double-counted, such as by also including a similar time-at-risk enabling condition or an IPL such as a procedural control that requires good housekeeping preventing the accumulation of dust that could form the explosible cloud.

3.3 Probability of Ignition or Initiation

The LOPA conditional probability of a flammable vapor, explosible dust cloud or combustible mist igniting or an uncontrolled reaction (such as an explosive decomposition) initiating is treated in various ways by different companies. The easiest way of treating this factor is to always assume an ignition or initiation probability of 100%, and have risk criteria that are appropriate to compare with this conservative approach. However, this can significantly overstate the risk in some cases, and it does not differentiate between scenarios where the probability of ignition is high versus those where it is likely to be quite low.

A comprehensive treatment of this topic is beyond the scope of this document, and typically requires expert evaluation. Simplified, rule-based approaches are commonly employed. For example, a published paper by Moosemiller (2009) reports the results of a project performed for the Explosion Research Cooperative that gives the following default probabilities for flammable vapor releases, based on a combination of published data and expert opinion:

0.15 = Probability of prompt ignition

0.30 = Probability of delayed ignition, given no prompt ignition

Since, as reported in the same paper, these values are highly dependent on a wide variety of material, release and environmental conditions, they should be replaced if at all possible with values that are supported by a study of the specific scenario conditions such as the temperature of the material being released as compared to its autoignition temperature, the minimum ignition energy of the material, and the release quantity, rate, duration, pressure, velocity and location (e.g., indoor vs. outdoor; proximity of continuous or intermittent ignition sources). The same paper gives correlations based on these factors.

A probability of ignition or initiation may be associated with three basic scenario types:

- Igniting a flammable or explosible atmosphere inside process equipment, resulting in an internal combustion reaction (usually a deflagration; transition to a detonation may be possible under the right conditions) that may or may not breach the primary containment.

- Igniting flammable vapors or an ignitable dust cloud external to process equipment, resulting in a flash fire, pool fire, jet fire, fireball and/or vapor cloud explosion.
- Imparting sufficient energy to a reactive material or mixture to initiate an uncontrolled reaction such as explosive decomposition or runaway polymerization.

These will each be treated in turn.

Probability of Ignition Inside Process Equipment

An example of an incident involving ignition inside process equipment occurred at the ARCO Chemical Company in Channelview Texas (CSB 2002).

On July 5, 1990, seventeen workers were killed when a 900,000 gal hydrocarbon-containing wastewater storage tank exploded. Although the tank had a nitrogen purge system, an investigation determined that the nitrogen purge failed, allowing a flammable atmosphere to develop in the vapor space of the tank. On the day of the incident, the tank was well above design specifications with respect to the amount of hydrocarbons present. In addition, an oxygen analyzer malfunctioned that could have provided warning of a flammable atmosphere inside the tank. While personnel were responding to correct the loss of nitrogen purge, ignition of flammable vapors inside the tank resulted in a catastrophic tank rupture. Although a number of possible ignition sources were identified, the actual source of ignition was never determined.

Scenarios involving ignition inside process equipment must be evaluated on a case-by-case basis. Some situations where use of a conditional modifier probability for internal ignition is not warranted include the following:

- When there is normally a flammable or explosible atmosphere inside the process equipment, and an internal deflagration will be initiated as soon as a sufficiently energetic ignition source is present, then the presence of the ignition source is the initiating event. The sufficiently energetic ignition source might be a result of a failed design feature such as a grounding and bonding system.
- When the presence of a flammable or explosible atmosphere and the presence of an ignition source are both short-duration abnormal events (i.e., revealed failures), then the concurrence of the hazardous atmosphere and the ignition source may need to be evaluated using the approach described in Appendix A for concurrent failures.

- For an initiating event that is a mechanical failure such as a screw conveyor failure, where the mechanical failure would likely result in sufficient energy being imparted to a flammable or explosible atmosphere to provide the required energy of ignition, then the ignition probability is often taken as 100%.

In situations where a conditional modifier is considered for internal ignition, the LOPA team should look at the entire scenario and all other factors involved before evaluating a probability of ignition.

When assessing a probability of internal ignition, it is advised that a person highly knowledgeable in this area be included in the evaluation to help decide upon a probability of ignition less than 100%. The following is a partial list of what may need to be considered in the evaluation:

- Historical experience
- Electrical classification inside enclosure
- Possibility of failed, incorrect or improperly installed internal components or instrumentation
- Static electricity generation (e.g., by streaming currents, free-falling of liquids, agitation, powder movement, processing of low-conductivity materials)
- Charge accumulation on ungrounded conductors (including consideration of grounding/bonding failures and of humidity control)
- Spark gap presence under normal or abnormal conditions
- Stray currents
- Vapor connections to remote locations where ignition sources may be present (e.g., thermal oxidizers, flares, vacuum pumps, vacuum trucks)
- Heat of adsorption (e.g., carbon beds or canisters)
- Internal and external surface temperatures
- Low autoignition temperature materials
- Low minimum ignition energy materials
- Pyrophoric, auto-oxidizing or catalytic materials (some may accumulate over time inside equipment)
- Internal normal or abnormal chemical reactions
- Hot work, including on adjacent equipment
- Rotating equipment, bearings, seals; friction, rubbing, impacts, vibration
- Adiabatic compression

Probability of Ignition Outside Process Equipment

Many of the same internal ignition considerations also apply to ignition outside of process equipment, such as following an atmospheric release or a loss of primary containment (LOPC) event. Table 3.1 in the preceding section on probability of hazardous atmosphere shows an example of probability of ignition as a conditional modifier.

In addition, other ignition mechanisms need to be assessed, such as:

- Vehicle/forklift traffic or equipment operation in the vicinity
- Drifting of flammable vapors to an ignition source such as hot work, hot surfaces or a compressor in an adjacent operating area
- Flashback from flammable vapors drawn into a building, fired heater or other ignition location
- Use of cell phones or other non-classified handheld devices in the area
- Use of portable or general-duty electrical equipment or tools in the area
- Smoking
- Lightning / electrical storms
- Static electricity generation and discharge such as from moving conveyors, powder handling or human activities in the area
- Improperly designed, installed or maintained electrical equipment within an electrically classified area.

Other factors needing to be considered when assessing the probability of ignition include:

- Size of the flammable vapor cloud or ignitable dust cloud (larger releases may extend beyond electrical classification boundaries and involve more potential ignition sources)
- Duration of the release (since probability of ignition by, e.g., vehicle traffic will increase as the duration of the release is extended)
- Overall number and strength of potential ignition sources
- Minimum ignition energy of the flammable vapors or ignitable dust

Guidelines for Engineering Design for Process Safety (CCPS 2012a) presents means of controlling ignition sources and thus reducing the probability of ignition.

CAUTION:
If a release occurs inside an electrically classified area, this is not by itself sufficient to give risk reduction credit for probability of ignition as a conditional modifier.
While establishing and maintaining electrically classified boundaries and equipment is good practice and may be required by codes, it does not guarantee the absence of a source of ignition, and large releases can be expected to extend beyond the electrically classified area boundary.

Immediate vs. delayed ignition. Additional fire/explosion mechanisms apply to releases outside of process equipment. Depending on the actual conditions, ignition of a flammable vapor cloud or ignitable dust cloud may result in a flash fire, pool fire, jet fire, fireball and/or vapor cloud explosion. Since a vapor cloud explosion generally requires delayed ignition, the probability of ignition conditional modifier may need to be broken down into the probability of immediate ignition vs. delayed ignition, as illustrated in the event tree of Figure 3.1. (See Section 3.4 for a discussion of vapor cloud explosion conditional modifier probabilities.)

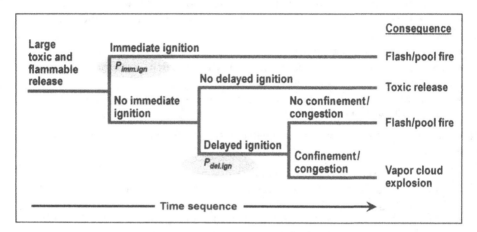

Figure 3.1 Probability of ignition conditional modifiers for external release.

Probability of Uncontrolled Reaction Initiation

The probability of initiating an explosive decomposition reaction in a condensed-phase material such as trinitrotoluene (TNT), picric acid or nitroglycerine is fairly straightforward. First, different mechanisms for imparting the necessary energy input to initiate the decomposition reaction are examined and characterized, including the potential effect of contamination. These mechanisms typically include shock or impact, friction, electrostatic discharge and heat. (Some unstable materials can be initiated by other mechanisms as well, such as by light.) Next, the available energy input is compared to what is known to be required to initiate the decomposition reaction. The energy input required is typically not a fixed value, but rather a range that can be characterized by a probability curve. The probability determined from the curve can then be used as a conditional modifier for the explosive decomposition scenario.

Conditional modifiers may not be appropriate to express the probability of initiating the reaction for many types of uncontrolled chemical reactions. Uncontrolled reactions are often treated on a deterministic basis. For example, if a maximum safe storage temperature is exceeded for a sufficient length of time or if a catalytic contaminant or inadequate inhibitor is present in a monomer storage tank, then an uncontrolled reaction is likely to result.

3.4 Probability of Explosion

This term is used for conditional modifiers where some type of explosion is possible, but would not be always expected to result. Since there are many explosion mechanisms, "probability of explosion" could take on any of several different meanings. Some common explosion mechanisms are discussed in this section.

Probability of Dust Explosion

Five basic factors must all be present for a dust explosion to occur:

- Confinement (usually provided either by a low-pressure enclosure such as a dust collector; or by the floor, walls and ceiling or roof of a room or building).
- Sufficient accumulation of an explosible dust within the confined volume (or part of the confined area, in some situations) to be above the lower explosible limit (and below the upper explosible limit, if applicable) of the specific dust within the confined volume, including consideration of the particle size distribution of the dust.
- Dispersion of the dust within the confined volume.
- A sufficiently energetic ignition source at the same time and location as the dispersed dust cloud (or that would be able to flash back to the dispersed dust cloud).
- Sufficient atmospheric oxygen or other oxidizer for combustion of the dispersed dust cloud.

Other factors may also affect the explosibility of the dust cloud, including the humidity, temperature and pressure. Note that many severe dust explosion incidents have involved more than one dust cloud dispersing and igniting, such as from accumulated dust in a building or area being dispersed by an initial explosion and resulting in a devastating secondary explosion.

Table 3.2 is an example from an actual incident illustrating the difficulties in evaluating the factors contributing to a dust explosion incident. Due to the number of possible factors involved in dust explosions, LOPA may not be the best method to use for analyzing potential dust explosion scenarios. Alternatives include conducting a Checklist Analysis for standard processes where prescriptive requirements can be followed to avoid dust explosions, or having a Fault Tree Analysis performed where a more detailed review is warranted.

Probability of ignition for explosible dust clouds can be treated in the same way as discussed in Section 3.3. Assuming the normal presence of sufficient atmospheric oxygen to support combustion, candidates for a "probability of explosion" conditional modifier for dust explosions would then include one or more of the following:

- Probability of confinement. (Without confinement, a flash fire would occur if an explosible dust cloud ignites, but confinement is necessary for a dust explosion to result.)
- Probability of sufficient dust accumulation of the explosible dust within the confined volume. This might be evaluated by comparison to criteria such as in NFPA 654 for what is considered sufficient dust accumulation to allow formation of an explosible dust cloud. This should not be double-counted if housekeeping to avoid dust accumulation has been credited elsewhere in the LOPA.
- Probability of a mechanism being present to disperse the dust within the confined volume. Note that this is often provided by a smaller initial explosion, dispersing the dust within a larger volume and resulting in a much more damaging secondary explosion.

Table 3.2 Imperial Sugar dust explosion incident causes (CSB 2009)

1	Sugar and cornstarch conveying equipment was not designed or maintained to minimize the release of sugar and sugar dust into the work area.
2	Inadequate housekeeping practices resulted in significant accumulations of combustible sugar and sugar dust on the floors and elevated surfaces throughout the packing buildings.
3	Airborne combustible sugar dust accumulated above the minimum explosible concentration inside the newly enclosed steel belt assembly under silos 1 and 2.
4	An overheated bearing in the steel belt conveyor most likely ignited a primary dust explosion.
5	The primary dust explosion inside the enclosed steel conveyor belt under silos 1 and 2 led to massive secondary dust explosions and fires throughout the packing buildings.
6	The 14 fatalities were most likely the result of the secondary explosions and fires.
7	Imperial Sugar emergency evacuation plans were inadequate. Emergency evacuation drills were not conducted, and prompt worker notification to evacuate in the event of an emergency was inadequate.

The following considerations also apply to an evaluation of dust explosions using conditional modifiers:

- One of the five factors that are required for a dust explosion will be the initiating event, in which case it could not be a conditional modifier. Conditional modifiers are generally associated with latent conditions; for example, the probability of sufficient dust accumulation, perhaps undetected due to being in a hidden space such as above a dropped ceiling, prior to a dispersion and ignition mechanism being present.

- The probability of explosion must meet the definition of a conditional modifier, and relate to the condition of the facility at a particular point or period of time in an incident sequence. The conditional modifier probability is the fraction of time the facility is in that condition when the incident sequence arrives at the pertinent step in the sequence.
- A possibility exists of a common mode failure between two or more of these factors, such as the cause for the dust dispersing within the confined volume also providing a source of ignition.
- If the dust explosion scenario is well-understood, other factors such as the relative humidity being sufficiently low or the particle size distribution being in the proper range may also have less than a 100% probability and be possibly evaluated as a conditional modifier.

Evaluation of dust explosion scenarios should be performed carefully with the input of knowledgeable persons on the assessment team. Evaluation of the scenarios in a quantitative manner can be very difficult and may require multiple mechanisms and scenarios to be assessed.

Probability of Vessel Rupture Explosion

The probability of internal vessel overpressurization exceeding the ultimate strength of the vessel may have a probabilistic aspect to it that can be treated using a conditional modifier. An example is shown in Table 3.3 where the likelihood of realizing the overpressurization consequence severity is treated on a probabilistic basis, perhaps by comparing a distribution of possible source pressures with a distribution of ultimate vessel strengths and estimating the overlap of the two distributions.

Some LOPAs would treat this on more of a deterministic basis, by comparing the maximum expected internal pressure to the MAWP to determine the expected result, which may result in a lesser consequence severity rather than a lower likelihood of the worst-case vessel rupture. In any case, some justification would be needed for crediting such a conditional modifier as a risk reduction factor, including verifying assumptions such as the vessel is properly designed, fit for duty, inspected and maintained.

Another example of a candidate scenario for employing a probability of vessel rupture explosion conditional modifier is ignition of a flammable vapor/air mixture inside a pressure vessel, where the vessel would generally be capable of containing the peak deflagration pressure if ignition occurred while the vessel internal pressure was near ambient. However, if the initial pressure at the time of ignition was elevated, the peak deflagration pressure could exceed the ultimate strength of the vessel and a vessel rupture explosion would result (unless an emergency relief device designed to protect against this particular scenario was installed and maintained and provided adequate protection on demand). If the initial process pressure at the time of ignition varies, then a conditional modifier

Table 3.3 Probability of explosion conditional modifier example

Scenario title: Breakthrough of a high-pressure fluid overpressurizes storage vessel TK-09; vessel rupture with ensuing pressure burst and liquid/spray/mist release to the surroundings.		Frequency	Probability
...			
Initiating event	Pressure letdown valve fails open, allowing high-pressure fluid into TK-09	0.1/yr	
Enabling condition	None		1
Conditional modifiers	Probability that the source pressure of upstream high-pressure fluid is sufficient to result in high enough pressure in TK-09 to exceed ultimate strength of the storage vessel when initiating event occurs		0.1
Scenario frequency without IPLs		0.01/yr	
Independent protection layer PFDs			
Emergency relief protection (PSV)			0.01
Total PFD for all IPLs			0.01
Scenario frequency with IPLs		0.0001/yr	

might be the fraction of time the initial process pressure was high enough that the peak deflagration pressure would exceed the ultimate strength of the vessel.

Other possibilities for conditional modifiers related to probability of vessel rupture explosion include dropping below a vessel's minimum design metal temperature resulting in brittle fracture, or exceeding a vessel's design temperature. Such scenarios require very careful evaluation by knowledgeable experts, and may involve calculations such as by the Larson-Miller method (API Std 530) or Omega methods (API RP 579) for creep rupture, taking into consideration additional factors such as corrosion rate when warranted.

Probability of Vapor Cloud Explosion

The probability that delayed ignition of a flammable vapor cloud will result in a vapor cloud explosion that generates a potentially damaging overpressure is the final branch point in the event tree of Figure 3.1 shown in the preceding section. It is generally recognized that a vapor cloud explosion requires some degree of confinement and/or congestion to develop. A probability-of-explosion conditional modifier might be used to indicate the fraction of time a flammable vapor cloud would find a confined/congested area prior to ignition.

For example, assume a flammable vapor cloud is formed due to a vapor or flashing liquid release from an isolated pipeline flange. Also assume there is no confinement/congestion in the immediate area of the release, and there is a confined/congested area in only one direction from the flange location. In this case, the probability of the wind blowing from the flange location to the confined/congested area could be used as a conditional modifier probability, based on actual meteorological wind direction data that is applicable to the plant location.

Two cautions are warranted with respect to a probability of vapor cloud explosion using a wind direction probability:

- Cloud slumping and terrain effects may need to be considered if the flammable vapor cloud is heavier than air.
- It should be ensured that the wind direction probability is determined in the context of the whole scenario being evaluated. For example, if the loss-of-containment cause is associated with maintenance turnarounds and the plant always has its turnarounds in the spring of the year, then the wind direction data need to apply to that time of year. (Note that this type of modifier should be subject to management of change if the modifier decreases the risk of the event. If a turnaround is taken in the fall, contrary to the normal practice, and taking the turnaround in spring decreased the required IPLs that were required due to the weather conditions, then the management of change process should be robust enough to identify that this change is significant and that compensating measures may be necessary.)

Note that this conditional modifier may be more likely to be used in a QRA than in a LOPA, depending on the approach used in the LOPA.

Probability of Deflagration-to-Detonation Transition

Situations involving ignition of near-stoichiometric, highly reactive and/or oxygen-enriched vapor mixtures have the possibility of the flame speed increasing to above sonic velocity and thus transitioning from a deflagration to a detonation. This may result from mechanisms such as pressure piling in equipment such as a pipeline having a large length-to-diameter ratio, or turbulence caused by pathway obstacles in confined/congested areas or enclosures. Even though the system has sufficient strength to contain the peak pressure of a deflagration, it is likely to fail under detonation conditions.

If there is a probabilistic aspect to the factors required for deflagration-to-detonation transition (DDT) to occur, such as a variation in the vapor composition, initial pressure, or length-to-diameter ratio of different parts of the process, then the fraction of situations where DDT can be anticipated is a candidate for a probability-of-explosion conditional modifier. As with other possible conditional modifiers described earlier, one or more knowledgeable persons need to be included in the LOPA to properly evaluate the conditional probability.

Figure 3.2 shows an example correlation that could be useful for assessing a conditional modifier for the probability of deflagration-to-detonation transition. Figure 3.2 correlates a distance parameter, which is twice the dimension of an ethylene cloud within its DDT composition limits of 5.9 to 9.3 percent by volume, versus a DDT conditional probability. For example, if dispersion modeling indicated the distance of the cloud within the 5.9 to 9.3% range was 10 ft, then the distance parameter would be equal to 20 ft. The corresponding probability of a transition to detonation occurring upon ignition of the cloud is predicted to be 0.5, or 50%. Note that Figure 3.2 is based on an extrapolation from hydrogen data, and does not explicitly take other factors such as degree of turbulence into account.

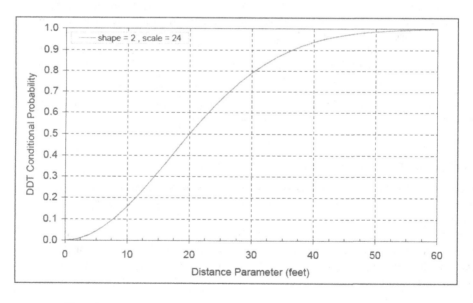

Figure 3.2 Example correlation for probability of ethylene DDT.
(Courtesy of Chevron Phillips Chemical, used with permission)

3.5 Probability of Personnel Presence

Probability of personnel presence is a conditional modifier that relates to the fraction of time people are likely to be <u>within the affected area</u> (sometimes termed *effect area* or *impact zone*) when a loss event occurs. Examples of conditional modifiers for the probability of personnel presence were given in Section 3.1 when defining and characterizing conditional modifiers.

Some key considerations for using (or not using) probability of personnel presence as a conditional modifier are as follows:

- This conditional modifier is often used in conjunction with the "probability of injury or fatality" conditional modifier described in the next section. It should be ensured that they are properly understood and interfaced such that they are not double-counting the same factor.

- In order to use these conditional modifiers, a more detailed view of how an incident sequence would progress once containment has been breached will often be required. The use of source term calculations and dispersion modeling, or of event trees such as the one shown in the preceding Figure 3.1, may be helpful in this regard. One word of caution: there should be a balance between the level of detail in the consequence estimation within a LOPA and the level of detail in the likelihood estimation.

- The probability of personnel presence should account for all personnel in the effect area, including routine operations, transient or short-term operations such as startups, maintenance work, anticipated abnormal situations and periods of time in which a larger group of people may be present. Personnel in adjacent units will also need to be considered if the event is large enough to affect more than the immediate area. The effectiveness of access control to potentially affected areas is another consideration. Note that scenarios involving startups and turnarounds will often need to be evaluated separately from continuous operation.

- The probability of personnel presence must be independent of the scenario being evaluated. This is often not the case when operator response to an alarm is involved, since part of the operator response may be to check out the situation in the field. (Note that this consideration will depend on the timing of the event sequence.) Other examples are where the scenario results in an abnormal noise (such as a high-pressure gas release, cavitating pump or splashing liquid), visual observation (such as an obscuring plume or venting vapor) or odor that personnel may get closer to in order to diagnose and attempt to correct.

- This aspect of the probability of personnel presence impacts procedures and training, by determining how operators are supposed to respond to events and whether their response actions will put personnel in harm's way.

- Another example where personnel presence should not be credited as a conditional modifier is when the presence of an operator would be required for the initiating event to occur, such as manual mischarging to a batch reactor, the opening of a wrong valve with short-term consequences such as draining to the floor, or an error made during a loading or unloading operation.
- Personnel presence should not be credited if it is included in the determination of the severity of consequences.
- Likewise, it should not be used if the company uses LOPA risk criteria that does not account for the presence of people (i.e., effectively assumes someone is present at any location 24 hrs/day, 365 days/year).
- The same approach can be used for employing a conditional modifier for the presence of people, valuable equipment, animals or other targets of concern in an appropriately determined effect area for establishing off-site, property damage, environmental or other impacts. An example would be if an effect area extends off-site to an agricultural field or a parking lot that is vacant most of the time. Care must be taken when using such a conditional modifier to ensure its ongoing validity, since property usage can change significantly over time. Unless robust management-of-change and revalidation procedures keep its LOPAs up to date, an organization may choose to limit usage of this conditional modifier to only on-site personnel presence; others may allow a broader usage. Even when limited to onsite personnel, the use of probability of personnel presence as a conditional modifier can expand the scope of the traditional MOC process.

An alternative to considering personnel presence as a conditional modifier is to consider it as an independent protection layer, but only if (a) administrative and/or engineering controls are in place to limit occupancy such as to specifically restrict exposure duration within an effect area to e.g. one person for no more than 10% or 1% of the time, and (b) the control of occupancy meets all of a company's LOPA IPL criteria for procedure-based control. For example, a LOPA might be evaluating a scenario associated with startup of a process unit where personnel are excluded from the process area during startup. In this case, the probability of personnel presence would represent the probability of not effectively controlling access to the area during startup.

> ### *CAUTION*
>
> *If exclusion of personnel is used as an independent protection layer,*
>
> *make sure the same factor is not double-counted as a personnel presence conditional modifier.*
>
> For example, during startup of a particular unit, the risk of a hazardous event occurring is assessed by the facility's management to be higher than during continuous operation. To reduce the likelihood of injury to personnel from a localized event during this period of startup, an operating procedure is put in place to minimize the probability of personnel presence. Prior to starting this operation, all work permits in the area are rescinded and all non-operations personnel are cleared from the area. Operators do a walk-through of the area to ensure that all personnel have been cleared. Temporary signage is put in place to indicate that the facility is in a hazardous startup mode and entry is not permitted. The operations personnel move into a protected area, announcements are made over the public address system, and only then is the higher-risk startup phase initiated.
>
> In this case, this operating procedure might be credited as an independent protection layer (IPL), with the probability of operators failing to perform the procedure or other personnel violating the procedure used as the IPL PFD. If this IPL is used in a LOPA, the probability of personnel presence should not be used as a conditional modifier in the same LOPA, since the two factors are not independent.

Determination of Effect Area

The effect area (impact zone) will vary depending on the specific event being analyzed. In most cases, its determination will require some form of consequence analysis, which can range from a quick worst-case estimate to detailed evaluations that may involve source term calculations, dispersion analysis, blast effect determinations, etc. Guidance is available from CCPS (1995, 1999, 2010) as well as other sources for the technical aspects of determining an effect area for various scenario types.

Deterministic calculations will require selection of appropriate fire, explosion and/or toxic release thresholds for determining the outer boundary of the effect area. These thresholds should correspond to the LOPA endpoints discussed in Section 1.4.

Dispersion modeling. For determining the effect area for outdoor toxic releases, dispersion modeling is generally required. Dispersion modeling may also be needed to determine the extent of a flammable cloud, such as for finding the area likely to be affected by a flash fire. Dispersion modeling generally requires calculation of the *source term* associated with the LOPA scenario being evaluated, which defines the composition of the material released to the surroundings as well as the release conditions of flow rate, duration, phase(s), pressure, temperature, location, elevation and orientation. This information is then provided as input to an appropriate dispersion model to determine the expected downwind distance and width of the toxic/flammable plume to the selected endpoint (such as the ERPG-2 concentration, LD_{LO} inhaled dose or 50% of the lower flammable limit), consistent with the risk criteria used as discussed in Section 1.4. The plume downwind distance and width will depend on the atmospheric stability and wind speed conditions selected for the analysis. *Guidelines for Use of Vapor Cloud Dispersion Models* (CCPS 1996) is one source of information on the details of dispersion modeling.

Toxic release effect areas. The effect area for outdoor toxic releases is generally elongated in the downwind direction, unless gravity slumping of a dense gas is the dominant effect. Calculation of the effect area to the selected endpoint can take a variety of forms depending on whether concentration or a combination of concentration and time is used as the endpoint, whether consideration is given to persons being indoors at the time of the release, and whether a different endpoint is used for on-site vs. off-site populations. Wind direction probabilities can be used based on meteorological data that is accurate for the actual scenario location.

Fire and explosion effect areas. For purposes such as determining an approximate effect area for a LOPA study, the effect distance for fire and explosion events is approximately equal in all directions, such that the effect area can be represented as a circle on a facility plot plan, unless the fire/explosion event has significant directionality such as a jet fire scenario or the splitting open of a pipe or vessel. The effect area boundary (endpoint) must be consistent with the facility's risk criteria.

Typical effect area boundaries used for flammability hazards include:

- Thermal radiation threshold, usually expressed as an incident thermal radiation heat flux for a specified time
- Lower flammable limit, or some percentage of the lower flammable limit
- Inside a diked area (e.g., for minor pool fires)
- Inside a room, enclosure or other area where escape is impeded and/or shielding is unavailable if the fire is an indoor scenario.

For explosion scenarios, a blast-effect endpoint such as a side-on overpressure of 0.5, 0.6 or 1 psi (3.4, 4.1 or 6.9 kPa) might be used, depending on the severity level being evaluated. Alternatives include a maximum expected missile/flying debris distance or a building damage level threshold.

Calculation of Personnel Presence Conditional Modifier Probability

The probability of personnel presence is generally calculated based on the fraction of time people are expected to be in the effect area for the specific scenario being evaluated. This is relatively straightforward if the effect being considered is only to one person, as presented in Table 3.4. The general approach is to:

1. Determine the boundaries of the effect area as described above.
2. Sum the total man-hours spent per day or per week within that effect area.
3. Use a calculation such as in Table 3.4 to determine the personnel presence conditional modifier probability.

Overlaying the effect area on an area plot plan, such as is illustrated in Figure 3.3, may be helpful when determining the probability of personnel presence. Differences in occupancies as a function of time of day will need to be factored into the man-hours determination. Note that persons within the effect area but inside a structure might be excluded from being counted in the man-hours total, if it is ensured that the structure is adequate to protect against the specific LOPA scenario being evaluated. Pertinent information such as occupancy data may be available from previously performed facility siting studies for the plant or process being examined.

Additional factors may need to be considered if more than one person can be in the effect area. For example, the severity categories being used by the company or facility conducting the study may have one severity category for a single serious or fatal injury and a higher severity category for multiple serious injuries or fatalities. In this case, the LOPA team may decide one or the other of these severity categories is most applicable to the scenario being evaluated and will then use a corresponding conditional modifier for probability of personnel presence. An alternative would be to consider a separate LOPA for each severity category, with the corresponding conditional modifier (s) pertaining to each severity category.

Table 3.4 Conditional modifier probability: One person present[1,2]

Time per week spent in effect area	Probability of personnel presence
17 hours	0.1
2 hours	0.01
10 minutes	0.001

[1] Total of all hours per week that any person may be in effect area, including all operations, maintenance, contractor, security, engineering and transient personnel, as well as persons present during peak periods such as breaks and assemblies.

[2] Caution: Only use this conditional modifier probability if the inclusion of conditional modifiers is reflected in the risk criteria pertaining to the facility being studied. Some companies might limit the probability of personnel presence conditional modifier value to be, for example:
- no less than 0.1, or
- no less than 0.1 without careful analysis and documentation of a lower probability, or
- no less than 0.01.

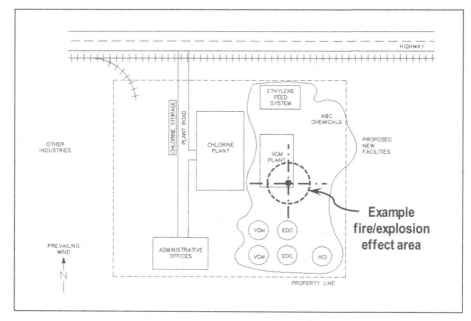

Figure 3.3 Example effect area for determining probability of personnel presence.

3.6 Probability of Injury or Fatality

A *probability of injury or fatality* conditional modifier relates to the probability that, given a person is within the effect area (impact zone) as determined in the preceding section, a serious injury or fatality would actually result.   This conditional modifier cannot be determined independently of the probability of personnel presence, since it will be affected by the endpoint chosen for calculating the effect area.

For example, for a given toxic release event, selecting the released material's ERPG-2 value for the toxic release endpoint will give a larger effect area than selecting the material's ERPG-3 or LC_{50}, thus likely resulting in a higher probability of personnel presence.   However, the average probability of serious injury or fatality within the larger ERPG-2 effect area will generally be lower than for the closer-in effect area of the ERPG-3 or LC_{50} endpoint.   Four approaches for defining the probability of injury or fatality conditional modifier in conjunction with the probability of personnel presence, with examples of each approach, are as follows.   In selecting the actual approach to use, the analyst should consider not only the specific scenario being evaluated but also the company's or facility's risk criteria basis.

Approach #1: Use P = 1 for the Probability of Injury or Fatality Conditional Modifier, Because of How Effect Area Boundary Is Defined

This approach is appropriate for impact mechanisms that have a relatively sharp boundary between little effect and serious effect, so that personnel present inside the effect area will have a near 100% probability of serious injury or fatality, whereas persons outside the effect area will have a low enough probability of serious injury or fatality that they can be assumed to be safe for purposes of the LOPA. Note that taking this approach is effectively the same as not using a conditional modifier for probability of serious injury or fatality. Loss events where this approach might be taken often include situations where the effect area boundary is a physical boundary such as a wall, dike, or berm; or it is the boundary of a flame front or a released solid or liquid material. Typical scenarios of this nature include:

- Flash fires. Persons inside the flash fire boundary are likely to be seriously or fatally affected by the fire thermal radiation and/or hot gases, but persons outside the flash fire boundary may receive first-degree or possibly second-degree burns to exposed skin; their thermal radiation exposure is generally brief enough to preclude serious injury or fatality.
- Small pool fires. Persons inside the diked area or other pool fire boundary are likely to be severely affected, but persons outside the boundary may be shielded or protected by the boundary itself.
- Hot material releases. Persons actually coming into contact with a significant release of the hot material can be severely burned, but injury is generally avoided if no contact occurs; examples include heat transfer fluids, steam, hot condensate and hot process fluid.

Approach #2: Apply the Same Probability of Injury or Fatality Conditional Modifier (P < 1) to Everyone Inside the Effect Area

This approach is appropriate for impact mechanisms that may have an effect that is both (a) relatively uniform across the effect area and (b) intermediate between an effect severe enough to cause injury or fatality to every person and an effect with little potential for severe injury. Scenarios of this nature might include:

- Building damage: If the persons inside a building are exposed to the same intermediate building damage level as a result of an external explosion. See Table 3.5 for an example of qualitative building damage descriptors that may be helpful in determining a conditional probability; see also CCPS 2012b for information on evaluating occupied buildings for the effects of external explosions.

- Indoor toxic exposure: If the indoor vapor concentration is relatively uniform and high enough to result in a dangerous inhaled concentration for a given duration of time, as compared against a probit curve or other means of assessing the conditional modifier probability of serious injury/fatality.

This approach can also be employed by calculating an average serious injury/fatality probability over all persons located within the effect area.

Approach #3: Estimate the Probability of Serious Injury or Fatality Conditional Modifier for Individual(S) Inside the Effect Area

When only one or very few persons are likely to be affected by a given loss event, the specific scenario can be studied to assess a probability of severe injury or fatality by the impact mechanisms applicable to the scenarios. (When more persons or off-site populations can be affected in a non-uniform way, a detailed analysis of probabilities for each person potentially affected is generally beyond the scope of the typical LOPA study.)   Some scenarios of this nature might include:

- Outdoor toxic exposure. Calculate a vapor concentration or inhaled dose at the personnel location and compare to a threshold value, toxic load or probit; or the conditional modifier probability may be the fraction of time the wind is blowing toward the population of concern.
- Shrapnel, fragments, flying debris. If personnel are not in the immediate area of an energetically failed piece of equipment, and a large area could be affected, the probability of impacting any particular point within the effect area can be small. Examples include compressor failure, blocked-in pump failure and vessel rupture explosion scenarios.

Gubinelli et al. (2004) reviewed the probability of a target being impacted by fragments generated by rupture of an overpressurized vessel. Impact probabilities calculated for a number of representative case studies were always less than 0.1, and were always less than 0.01 for secondary targets at distances greater than 50 m (164 ft). A simplified model such as this might be able to be employed in a LOPA when evaluating the probability of impacting a particular target population. Other blast effects also need to be taken into account, particularly if the target is close to the blast source.

Table 3.5 Antiterrorism building protection level vs. potential injury (DoD 2007)

Level of Protection	Potential Building Damage / Performance[1]	Potential Door and Glazing Hazards[2]	Potential Injury
Below Anti-terrorism Standards[3]	Severe damage. Progressive collapse likely. Space in and around damaged area will be unusable.	Doors and windows will fail catastrophically and result in lethal hazards. (High hazard rating)	Majority of personnel in collapse region suffer fatalities. Potential fatalities in areas outside of collapsed area likely.
Very Low	Heavy damage – Onset of structural collapse, but progressive collapse is unlikely. Space in and around damaged area will be unusable.	Glazing will fracture, come out of the frame and is likely to be propelled into the building, with the potential to cause serious injuries. (Low hazard rating) Doors may be propelled into rooms, presenting serious hazards.	Majority of personnel in damaged area suffer serious injuries with a potential for fatalities. Personnel in areas outside damaged area will experience minor to moderate injuries.
Low	Moderate damage – Building damage will not be economically repairable. Progressive collapse will not occur. Space in and around damaged area will be unusable.	Glazing will fracture, potentially come out of the frame, but at a reduced velocity; does not present a significant injury hazard. (Very low hazard rating) Doors may fail, but they will rebound out of their frames, presenting minimal hazards.	Majority of personnel in damaged area suffer minor to moderate injuries with the potential for a few serious injuries, but fatalities are unlikely. Personnel in areas outside damaged areas will potentially experience minor to moderate injuries.
Medium	Minor damage – Building damage will be economically repairable. Space in and around damaged area can be used and will be fully functional after cleanup and repairs.	Glazing will fracture, remain in the frame and results in a minimal hazard consisting of glass dust and slivers. (Minimal hazard rating) Doors will stay in frames but will not be reusable.	Personnel in damaged area potentially suffer minor to moderate injuries, but fatalities are unlikely. Personnel in areas outside damaged areas will potentially experience superficial injuries.
High	Minimal damage. No permanent deformations. The facility will be immediately operable.	Glazing will not break. (No hazard rating) Doors will be reusable.	Only superficial injuries are likely.

Notes:
[1] For damage/performance descriptions for primary, secondary and non-structural members, refer to UFC 4-020-02, DoD Security Engineering Facilities Design Manual.
[2] Glazing hazard levels are from ASTM F 1642. (Use of wire reinforcement or other protective glass design can significantly reduce glazing hazards.)
[3] This is not a level of protection, and should never be a design goal. It only defines a realm of more severe structural response, and may provide useful information in some cases.

Approach #4: Combined Approach

Evaluation of some scenarios may warrant a combination of the above approaches. For example, a scenario involving a vessel rupture explosion with an accompanying large fireball might be evaluated by determining a blast effects radius and/or thermal radiation radius within which the probability of serious injury/fatality is equal to 1, combined with an evaluation of the conditional modifier probability for one or more persons at specific locations outside this radius by flying debris and/or thermal radiation effects. This analysis would need to be carefully combined with the probability of personnel presence in each of these separate areas to have the LOPA properly determine the overall scenario risk.

Notes and Cautions

The following are further considerations when using a conditional modifier that assesses a probability of serious injury or fatality:

- More than one consequence may need to be evaluated. For example, an initial flash fire may need to be evaluated as one impact area, followed by a pool fire with a different impact area.
- The conditional modifier probability must apply for the entire effect area, not just at the outside boundary of the effect area where the intensity of the effect may be marginal.
- If personnel are not in the effect area at the time of the loss event, but might enter the effect area while the danger is present (such as for a carbon monoxide release that is invisible and odorless), then use of an engineered safeguard that would detect the dangerous condition and warn before entry could be considered an independent protection layer and not a conditional modifier.
- The characteristics of specific hazardous materials involved in a release scenario may need to be taken into account when evaluating the probability of injury or fatality. For example, some chemicals are severe lachrymators (tear-producers) or have other properties that would hinder escape and increase exposure time. Others such as titanium tetrachloride form dense clouds that greatly reduce visibility. On the other hand, chemicals such as carbon monoxide and dimethyl sulfate have little or no warning properties, thus likely increasing exposure time but in a different way.

Fully quantitative risk analyses may take additional, related conditional modifiers into account that would not generally be employed in a Layer of Protection Analysis, although one or more might be used by a company if included in its documented LOPA approach. Additional conditional modifiers related to the probability of serious injury or fatality can include:

- Probability of unsuccessful pre-evacuation or sheltering
- Probability of unsuccessful escape or shielding
- Probability of inadequate medical treatment or resuscitation.

These probabilities can be significantly different for on-site versus off-site potentially affected populations. More specific or unique hazards may have other possible conditional modifiers.

3.7 Probability of Equipment Damage or Other Financial Impact

Discussed in this section is a conditional modifier that may be appropriate for some scenarios when evaluating economic impacts such as property damage and business interruption costs. This factor represents the probability that a significant economic impact would result, regardless of whether any independent protection layers are present.

An example is shown in Table 3.6, in which liquid carryover into a compressor may result in only slight damage to the compressor (e.g., coupling failure) most of the time (perhaps based on historical experience), but there is a probability that the outcome would be major damage to the compressor. The conditional modifier gives the probability that a major economic impact would result, given liquid carryover to the compressor. This conditional modifier would represent the range of possible outcomes for the given scenario. It is a separate and distinct consideration from any independent protection layers that may be installed, such as if the compressor suction knockout pot had a level sensor to automatically and independently shut off the compressor on high-high level in the knockout pot. The outcome in an actual situation will depend on the compressor type, rate of liquid carryover, density of carried-over fluid and size of equipment. Input from knowledgeable persons would be desirable in assessing this conditional probability in an actual LOPA study.

Table 3.6 Probability of financial impact conditional modifier example

Scenario title: Condensate always present to lesser or greater degree in gas stream entering compressor. Knockout pot removing condensate has automatic drain-out level control. Level control fails, allowing condensate level to increase in knockout pot to the point of liquid carryover into the compressor. Major compressor damage requiring total rebuild could result, with associated damage costs, unscheduled downtime and loss of production. Breach of compressor housing not anticipated.

		Frequency	Probability
...			
Initiating event	Knockout pot automatic drain-out level control fails such that condensate is not drained out	0.1/yr	
Enabling condition	None		1
Conditional modifier	Probability of financial impact (probability of major compressor damage with associated business impacts, given liquid carryover into compressor, based on historical experience)		0.1
Frequency of unmitigated consequence		0.01/yr	
...			

Tables 3.7 and 3.8 give two examples of data sets that may be useful in determining a probability of financial impact conditional modifier. These tables present expected damage levels to various types of process equipment as a function of blast overpressure. The tables do not differentiate between side-on and reflected blast overpressures, and also do not indicate the blast duration or impulse. As such, they should be used with caution, and may be more useful for assessing relative damage levels than absolute values.

In addition to process equipment, the value of building damage can also be substantial. Building damage level evaluations go beyond most LOPAs and may need to consider blast impulse instead of overpressure, particularly for vapor cloud explosions.

Table 3.7 Explosion damage to equipment and structures (Swiss Re 1998)

Within the following blast rings	Effect
5 psi	80% damage to typical process equipment
	100% damage to sensitive structures (such as cooling towers)
2 psi	40% damage to typical process equipment
	80% damage to sensitive structures
1 psi	5% damage to typical process equipment
	40% damage to sensitive structures

Table 3.8 Explosion effects on processing equipment (Stephens 1970)

Equipment	Magnitude	Effect
Cone-roof storage tank	1 psi	Roof collapses
	3 psi	Half-filled tank uplifts
	6.5 psi	90% filled tank uplifts
Floating-roof storage tank	3 psi	Half-filled tank uplifts
	6.5 psi	90% filled tank uplifts
	20 psi	Roof collapses
Spherical storage tank	8 psi	Bracing fails
	14 psi	Unit moves and pipes fail
	16 psi	Unit overturns or is destroyed
Chemical reactor	2 psi	Onset of damage, gauges broken
	4 psi	Unit moves or pipes break
	6.5 psi	Frame deforms
	9 psi	Unit overturns or is destroyed
Regenerator	3 psi	Bracing fails
	5 psi	Unit moves, pipes break, frame deforms
	7.5 psi	Unit overturns or is destroyed
Catalytic cracking reactor	3.5 psi	Bracing fails
	7 psi	Unit moves and pipes break
	12 psi	Unit overturns or is destroyed
Horizontal pressure vessel	6 psi	Frame deforms, unit moves, and pipes break
	9 psi	Unit overturns or is destroyed
Vertical pressure vessel	12 psi	Unit moves and pipes break
	14 psi	Unit overturns or is destroyed
Cooling tower	0.2 to 0.5 psi	Louvers fail
	2 psi	Inner parts damaged
	3.5 psi	Frame collapses
Fired heater	2 psi	Brick cracks
	2.5 psi	Unit moves and pipes break
	5 psi	Unit overturns or is destroyed

Table 3.8 *(Continued)*

Equipment	Magnitude	Effect
Filter	2 psi	Debris / missile damage occurs
	4.5 psi	Inner parts are damaged
	9.5 psi	Unit moves on foundation
	12 psi	Unit overturns or is destroyed
Distillation (fractionation) column	5.5 psi	Frame cracks
	7 psi	Unit overturns or is destroyed
Extraction column	6.5 psi	Unit moves and pipes break
	10 psi	Unit moves on foundation
	12 psi	Unit overturns or is destroyed
Heat exchanger	7.5 psi	Unit moves and pipes break
	9 psi	Unit overturns or is destroyed
Steam turbine	7.5 psi	Unit moves and pipes break
	12 psi	Controls are damaged
	14 psi	Piping breaks
	20 psi	Unit moves on foundation
Pump	12 psi	Unit moves and pipes break
	16 psi	Unit moves on foundation
Blower	5 psi	Case is damaged
	10 psi	Unit overturns or is destroyed
Electric motor	5 psi	Debris / missile damage occurs
	9 psi	Power lines are severed
	20 psi	Unit moves on foundation
Electric transformer	4.5 psi	Debris / missile damage occurs
	7.5 psi	Power lines are severed
	10 psi	Unit overturns or is destroyed
Pipe supports	3.5 psi	Frame deforms
	6 psi	Piping breaks, frame collapses
Gas meter	4.5 psi	Case is damaged
Gas regulator	6 psi	Unit moves and pipes break
	10 psi	Controls and case are damaged

3.8 Documenting, Managing and Validating Conditional Modifiers

As was the case for enabling conditions, the examples shown in this chapter illustrate only one way of documenting conditional modifiers in LOPAs. Some companies require more than the conditional modifier description and probability in the LOPA documentation, such as source references or calculations to back up conditional modifier probabilities.

The documentation shows the justification for the conditional modifier probability selected and aids in auditability and ongoing process safety management practices such as management of change. The documentation also provides the basis for proving and/or maintaining the required reliability and effectiveness of the credited risk reduction measures.

Conditional modifiers used in a LOPA should invoke the same degree of administrative controls and rigor as that of the Process Safety Information (PSI) for the hazardous process unit under evaluation. The PSI requires proper documentation, with supporting calculations and auditable assumptions, which are maintained through the facility's management of change (MOC) process. Periodic revalidation is also warranted, such as in conjunction with the Process Hazard Analysis (PHA). Depending on the nature of the conditional modifiers, personnel involved in the operation of the process may need to be trained on the basis for conditional modifiers, such as limited personnel presence within a potential effect area. The use and maintenance of conditional modifiers in accordance with documented expectations would be an appropriate subject for compliance auditing in accordance with a facility's written management system.

As was discussed in Section 2.6 for enabling conditions, in addition to the documentation of conditional modifiers and their respective probabilities in a LOPA, an organization may incorporate a further step of validation to its basic LOPA approach to ensure that the selected conditional modifier values are appropriate. For example, if a LOPA is performed for an expansion project that includes loading of a flammable liquid product into railcars, assumptions might need to be made regarding how the product loading operation will be performed, as these assumptions could affect a "probability of hazardous atmosphere" conditional modifier. After the actual standard operating procedures are established, these assumptions can be validated and the LOPA risk calculations adjusted if necessary to reflect the actual procedure to be employed.

4
Application to Other Methods

Enabling conditions and conditional modifiers are used not only in Layer of Protection Analyses, but also in other hazard evaluation methodologies. These include methods that are both more detailed and quantitative than the typical LOPA (see Section 4.1 for their application in Quantitative Risk Analyses) and less detailed than the typical LOPA (see Section 4.2 for their application in combination with scenario identification methods such as HAZOP Studies). The application of conditional modifiers to more qualitative methods using barrier analysis and diagrams is discussed in Section 4.3.

4.1 Quantitative Risk Analysis

Enabling conditions and conditional modifiers were employed in quantitative risk analyses (QRAs) for many years before LOPA was developed. Their function has generally been to take into account factors that are not related to system failures, in order to have an improved risk estimate and eliminate unnecessary conservatism from the analysis. This section illustrates how enabling conditions and conditional modifiers might be used in the context of three QRA approaches: Fault Tree Analysis (FTA), Event Tree Analysis (ETA) and consequence analysis.

Fault Tree Analysis

Fault Tree Analysis (FTA) is a deductive logic modeling technique that traces a loss event ("Top event") back to the various combinations of equipment failures, human errors and other conditions required for the loss event to be realized. A fault tree might include enabling conditions and/or conditional modifiers as part of the logic model.

Figure 4.1 illustrates how an enabling condition might be included in a FTA, including quantification. In this example, the enabling condition is depicted using a "house" symbol, indicating it is a normal or expected event. It is assumed for purposes of this example that loss of steam supply is independent of low ambient temperature. (Note that this may not always be the case. For example, loss of steam supply could be caused by freeze-up of control valves or traps in cold weather.)

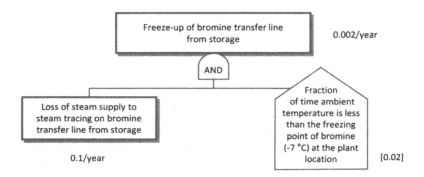

Figure 4.1 Illustration of enabling condition usage in a Fault Tree Analysis (quantification is for example purposes only).

Figure 4.2 shows the possible use of conditional modifiers in a fault tree. The first two conditional modifiers are depicted using an oval symbol, indicating they are not fault conditions but probabilistic factors. Ignition source presence would likely be developed further to identify specific sources. In both Figure 4.1 and Figure 4.2, the shading on some of the symbols point out the "initiating branch" events; i.e., those quantified with frequencies rather than probabilities.

Event Tree Analysis

Event Tree Analysis (ETA) is an inductive method that traces all possible outcomes of an initiating event based on the success or failure of possible barriers encountered in time sequence following the occurrence of the initiating event. Mitigation event trees studying the possible outcomes of a loss event may include one or more conditional modifiers as event tree branching conditions, as illustrated in Figure 4.3. The conditional modifier most likely to be encountered in event trees is probability of ignition. Enabling conditions are not generally used in event trees, since event trees focus on all possible outcomes assuming the starting event has already occurred.

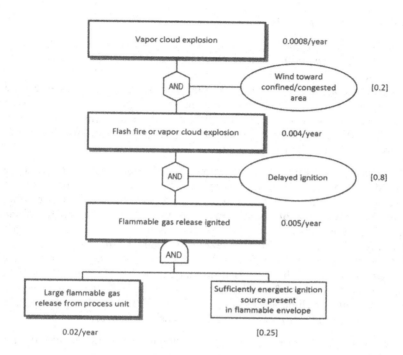

Figure 4.2 Illustration of conditional modifiers usage in a Fault Tree Analysis (quantification for example purposes only).

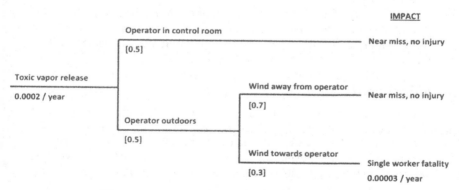

Figure 4.3 Event tree with conditional modifiers.

Quantitative Impact Analysis

The purpose of an *impact analysis* is to determine quantitatively the severity of consequences (i.e., the *impact*) of a loss event such as an explosion or a toxic gas release, regardless of the frequency of the loss event (CCPS 1995, CCPS 1999). This is usually done as part of a larger quantitative risk analysis. Impacts can be personal injury or fatality, environmental, business and/or property damage. Note that impact analyses will be taken to different endpoints (as described in Section 1.4) by different companies, or even within the same company, depending on the purpose of the quantitative risk analysis.

Conditional modifiers are commonly used along with consequence analyses to determine the quantitative impact. For example, the impact of a bursting vessel explosion event is likely to include consideration of personnel presence in the effect area of the explosion as well as an assessment of the probability of severe injury or fatality to persons in the effect area. That probability might be assessed for each explosion damage mechanism, such as direct blast overpressure, missiles/flying debris, translation (i.e., the blast wave throwing a person against a hard surface or sharp object) and startle effect (i.e., causing a person to lose his/her balance or otherwise fall from an elevated location). It may also include an assessment of the effect of building damage to persons inside an occupied structure within the explosion effect area.

For toxic gas or vapor releases, conditional modifiers quantified in a consequence analysis might include some or all of the following:

- The fraction of time environmental conditions (especially wind direction, wind speed and atmospheric stability) are such that the release plume will affect a receptor location (P_{env})
- The probability of no escape, evacuation or sheltering ($P_{no\ escape}$)
- A probit analysis of a calculated inhaled dose versus the probability of serious injury or fatality, or an estimate of the number of persons within the boundary of a concentration or dose threshold (P_{LD})
- The probability of unsuccessful medical treatment, given exposure to the toxic vapors ($P_{no\ resusc}$).

The consequence analysis may examine the expected effects under a full range of atmospheric conditions for each postulated release scenario, and may take into account other factors such as different impacts to persons who are outdoors vs. indoors. An example equation using the factors in the bulleted list above with the toxic release frequency $F_{release}$ to calculate a fatality frequency $F_{fatality}$ is as follows:

$$F_{fatality} = F_{release} \cdot P_{env} \cdot P_{no\ escape} \cdot P_{LD} \cdot P_{no\ resusc}$$

The analysis can be much more detailed than this simplified example, calculating a received dose at each receptor location under each environmental condition and taking into account other factors such as a distribution of ambient temperatures.

Note that the LOPA analyst can use this type of quantitative analysis when a more detailed evaluation needs to be made of scenarios such as simple loss-of-containment events that have few (if any) other types of risk reduction. The LOPA analyst should be aware of potential for common mode failures when multiple conditional modifiers are used, to avoid a nonconservative risk estimate.

4.2 Use of Enabling Conditions and Conditional Modifiers with Scenario Identification Methods

Any scenario-based hazard evaluation technique (CCPS 2008) can be used to identify potential incident scenarios that can serve as the starting point for a Layer of Protection Analysis (LOPA). Several of these methodologies can also be extended to include aspects of a LOPA such as are discussed in these *Guidelines*, either in the same team-based review or in a follow-up review. These aspects include:

- Estimating the scenario risk for scenarios exceeding a threshold consequence of concern or meeting other screening criteria.
- For each such scenario, evaluating the initiating cause (initiating event) frequency, consequence severity and effectiveness of IPLs on an order-of-magnitude basis using best-estimate and/or rule-based values.
- Including conditional modifiers and/or enabling conditions to estimate the likelihood of harm posed by the scenario, if their use is consistent with how the facility's risk criteria are established.
- Comparing the calculated scenario risk to a risk goal to determine the adequacy of existing risk control measures.

To be extended in this manner, the scenario development part of the hazard evaluation must be conducted using an approach that generates individual scenarios amenable to risk evaluation. Hazard evaluation techniques tending to fit this requirement include the following when documented in a tabular or database format:

- What-If Analysis
- What-If/Checklist Analysis
- Hazard and Operability (HAZOP) Studies
- Failure Modes and Effects Analysis (FMEA).

HAZOP Studies that have been performed in this manner are sometimes termed *HAZOP/LOPA Studies* or just *HAZOP/LOPAs*. Also, since What-If, What-If/Checklist and FMEA are inductive approaches that begin with a "what-if" condition or a failure mode that is generally an initiating cause or initiating event, they require no modification of the basic approach to employ the extensions listed above.

Note that HAZOP studies must use a cause-by-cause approach as described in CCPS (2008) rather than a deviation-by-deviation approach to employ all of the extensions listed above. Deviation-by-deviation HAZOP Studies do not associate independent protection layers with specific cause-consequence pairs. To employ the extensions listed above, the HAZOP Study must ensure the safeguard IPLs align with specific cause-consequence pairs to evaluate scenario risk as is done in LOPAs.

As is the case for LOPAs, not all of these extended hazard evaluations will include consideration of enabling conditions and/or conditional modifiers. Whether or not enabling conditions and/or conditional modifiers are used, the risk estimation approach employed must match how the tolerable risk level was established. It is common for companies to employ some aspects of these extended hazard evaluations, even if enabling conditions and conditional modifiers are not utilized. For example, initiating cause or unmitigated risk frequency scales based on orders of magnitude (1/yr, 0.1/yr, etc.) are in widespread use.

In addition, as discussed earlier, risks need to be managed at all levels, not just at the individual scenario level. Calculating scenario risks for a full set of process upset scenarios allows the possibility of aggregating scenario risks so they can be managed at the process or facility level as well as for individual scenarios (Johnson 2010). However, it should be noted that if screening-level frequencies and probabilities are used in a HAZOP/LOPA as is done in regular LOPAs, the conservatism built into the analysis may lead to a relatively high risk prediction when aggregated to the process level. The difference between risk criteria at single-scenario and process levels is discussed in *Guidelines for Developing Quantitative Safety Risk Criteria* (CCPS 2009) and further developed by Chastain (2010).

As compared to a common approach of conducting a qualitative hazard evaluation followed by LOPAs of selected scenarios, HAZOP/LOPAs and other extended scenario-based evaluations have both advantages and disadvantages (Johnson 2011). Possible advantages include:

- A LOPA-type evaluation is done by a full hazard evaluation team rather than by a LOPA analyst.
- The necessity for a separate LOPA may be precluded.
- LOPA standard values can be used, such as for initiating event frequencies and IPL PFDs.
- Documented scenario risk estimates can be obtained, as compared to using worst-case loss event impacts without conditional modifiers.
- Alignment is gained between hazard evaluations and LOPAs within an organization, by using the same rules and standard values.
- Other benefits common to LOPAs include the use of LOPA standard screening values and the possibility of comparing risk estimates among scenarios if the same method is employed.
    - Possible disadvantages HAZOP/LOPAs and other extended scenario-based evaluations include:

- More team review time is required than a qualitative hazard evaluation, and somewhat more than a detailed HAZOP Study using a risk matrix.
- The facilitator must be knowledgeable in both hazard evaluation and LOPA methods.

HAZOP/LOPA Example

Table 4.1 illustrates one way to document two HAZOP/LOPA scenarios having the same initiating cause. Each scenario includes two conditional modifiers (CM's) as risk-reduction factors. The cause frequency magnitude (F) of 0 represents a cause frequency of 10^0 events per year, or once a year. A risk reduction magnitude (RRM) of 1 represents one order of magnitude of risk reduction afforded by the independent protection layer and by each of the two conditional modifiers. The RRMs are subtracted from F to give an overall scenario likelihood magnitude (L) of -3, corresponding to a 10^{-3} per year outcome frequency. L is added to the consequence severity magnitudes (S) of 6 or 5 for the two scenarios to give the overall scenario risk magnitudes (R) of +3 and +2.

Adequacy of risk control measures can be assessed by comparing these scenario risk magnitudes to a risk magnitude goal, or by looking up the intersection of L and S on a risk matrix. If, for example, the risk goal for the company performing the HAZOP/LOPA in Table 4.1 is for every scenario to have a risk magnitude of +1 or less, then the risks posed by the first and second scenarios in Table 4.1 would need to be reduced by at least two and one order of magnitude, respectively.

Table 4.1 Example HAZOP/LOPA scenarios *(see text)*

Step 1. Charge waste treatment reactor with 5.3 m³ of aqueous cyanide-containing waste.								
GUIDE WORD, Deviation	Cause	F	Consequences	S	Safeguards & Risk-Reduction Factors	RRM	Scenario L	R
MORE > 5.3 m³ of aqueous cyanide-containing waste into treatment reactor	Operator performs waste addition step twice	0	Cyanide concentration of 0.6% at oxidation step; HCN generated; release of HCN gas from scrubber stack; potential for on-site or off-site persons to inhale HCN; fatalities or multiple severe injuries possible	6	IPL: Independent verification check in procedure of total quantity added before proceeding	1	-3	+3
					CM: Person(s) are unlikely to be in HCN release plume effect area (dominant wind is away from populated on-site locations; no expected correlation between scenario and personnel presence)	1		
					CM: Fatalities or multiple serious injuries are unlikely, given person(s) are in effect area of HCN plume (limited vapor concentration; on-site medical treatment available)	1		
MORE > 5.3 m³ of aqueous cyanide-containing waste into treatment reactor	Operator performs waste addition step twice	0	Cyanide concentration of 0.6% at oxidation step; HCN generated; release of HCN gas from scrubber stack; ignition of HCN flammable vapors; personnel exposure to flash fire thermal radiation; moderate to severe burns possible	5	IPL: Independent verification check in procedure of total quantity added before proceeding	1	-3	+2
					CM: Ignition source control (Class I Div 2 area; no traffic or continuous ignition sources nearby; hot work not performed while process is operating)	1		
					CM: Person(s) unlikely to be in HCN flash fire effect area (flammable boundary is very limited; no work assignments near scrubber stack; no expected correlation between scenario and personnel presence)	1		

Application to Other Methods

Although less commonly done, it is possible to employ the same order-of-magnitude LOPA-type scenario analysis approach as above to the other scenario-based hazard evaluation techniques listed earlier (What-If Analysis, What/If-Checklist, FMEA). However, for these methods, it is important to distinguish between those failure modes or what-if questions that are initiating causes and those that pertain to other conditions such as unrevealed failures in standby safety systems. For example, if evaluating the failure modes and effects of a relief device on a pressurized system, the failure mode of premature opening of the relief device is an initiating cause that can be evaluated as to its likelihood, severity and independent protection layers. However, another failure mode of the same component, such as the relief device failing stuck closed, can be an unrevealed failure. The pressurized system can still operate normally with the relief device failed in this mode. Some other system failure is necessary to occur as an initiating cause of a high pressure deviation, in order for the relief device failure to be part of a scenario risk evaluation. Hence, the failure mode of a relief device failing stuck closed would not be developed further as a scenario.

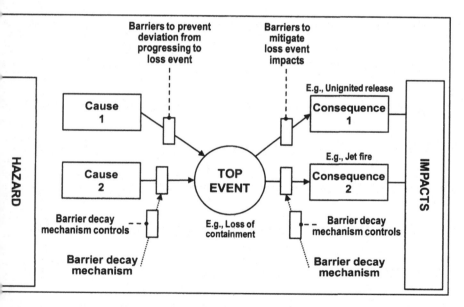

Figure 4.4 Bow-tie diagram concept (after Pitblado and Tahilramani 2009, using terminology in CCPS 2008)

4.3 Barrier Analysis and Diagrams

Conditional modifiers may also find their way into more qualitative methods such as barrier analysis techniques. One such technique uses "bow-tie" diagrams, as illustrated in Figure 4.4, to graphically depict the preventive and mitigative "safety barriers" that are in place to protect against a specific loss event such as a major loss of containment event. These safety barriers might include the hardware and/or administrative controls associated with some conditional modifiers such as ignition source control and the presence of personnel in the potential effect area of a loss event.

The bow-tie loss event is shown in Figure 4.4 as a "Top event," using Fault Tree Analysis terminology. Although Figure 4.4 only shows one preventive and one mitigative safety barrier for each cause and consequence, there are often multiple preventive barriers and multiple mitigative barriers to protect against major loss events.

Appendices

A – Simultaneous Failures and "Double Jeopardy"

B – Peak Risk Concepts

C – Example Rule Set for LOPA Enabling Conditions

Appendix A
Simultaneous Failures and "Double Jeopardy"

Included in this Appendix is a discussion of when it is and when it is not appropriate to consider simultaneous failures in the risk analysis of potential incident sequences. This issue sometimes arises when considering enabling conditions and conditional modifiers, since LOPAs that include them have multiple factors to combine when calculating scenario risk.

Considering simultaneous failures is often characterized as "double jeopardy" and thereby disallowed. However, in a limited number of situations, it is valid and even necessary to analyze the possible occurrence of two or more concurrent failures, as explained in the following paragraphs.

What is "Double Jeopardy"?

In the context of risk analysis, "double jeopardy" has the basic meaning of the occurrence of two simultaneous failures in a system. As is discussed below, "double jeopardy" can be more precisely defined as the *concurrent incidence of two independent initiating events or other revealed failures.* (The frequency of two independent failures or human errors occurring at exactly the same time is essentially zero, so that it is more proper to speak of *concurrent* failures rather than *simultaneous* failures. One device or system will always fail before the other, so the failures will never be truly simultaneous.)

Constitutional or statutory protections in various countries forbid a person from being tried in court twice for the same offense, which is popularly known as "double jeopardy." By extension, the connotation of calling something "double jeopardy" in a hazard evaluation or risk analysis is that it is disallowed from being considered further, by whatever rules or conventions are being used for the analysis. However, this term is often abused or not used with a proper understanding of the different types of concurrent failures.

What "Double Jeopardy" is NOT

An important distinction must be made between the occurrence of two concurrent system failures and the occurrence of a failure while another part of the system is already in the failed state or a state that would otherwise allow an incident sequence to proceed to a loss event. The following situations are **not** "double jeopardy":

- The consideration of two or more *unrevealed (latent)* failures.
- The consideration of two or more failures or conditions when only one is a *revealed* failure and the remainder are *unrevealed* failures, failures that are *tied to the occurrence of the revealed failure* and/or system states or conditions such as enabling conditions and conditional modifiers.

It is obvious that the distinction between *revealed* and *unrevealed* failures needs to be drawn for these situations to be properly understood.

Revealed vs. Unrevealed Failures

A *revealed* failure (sometimes termed an *announced* failure) is one that is readily apparent by its effect on the system or surroundings, such as a failure of the primary containment system or a control system failure that rapidly results in a process deviation such as loss of flow or high system pressure. Revealed failures are characterized by having a relatively short duration before either being detected or causing a loss event; i.e., the failure condition is manifested within seconds to hours. This short duration will be an important consideration when evaluating concurrent failures, as described later in this Appendix.

By contrast, an *unrevealed* or *latent* failure (sometimes termed and *unannounced* failure) does not have an immediate effect on the system or surroundings, so it can persist in the failed state for an extended duration (often months or years) before being detected and corrected or before being needed to perform its intended function and not doing so. Examples of unrevealed failures include such situations as the inlet to a relief device being plugged, an emergency block valve being stuck in the open position, or a high-high level float switch having holes corroded in its float device. In each case, the system can continue to operate indefinitely until the failure is detected by routine inspections, testing or maintenance or until the device is called upon to perform a protective function and fails to do so. This highlights the importance of periodically performing functional tests of standby safety systems.

In a typical incident sequence, revealed failures are most closely associated with initiating events that result in system deviations (abnormal situations) or loss events. Unrevealed failures are generally associated with weaknesses or "holes" in layers of protection, as in the "Swiss cheese" model of Figure A.1. In this figure, the initiating event frequency can be considered as the frequency of challenges to the protective barriers. The arrow demonstrates what would happen if the initiating event occurred and protective barriers (IPLs) were not in place.

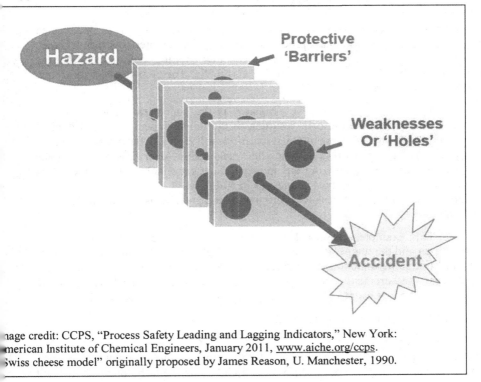

Image credit: CCPS, "Process Safety Leading and Lagging Indicators," New York: American Institute of Chemical Engineers, January 2011, www.aiche.org/ccps. "Swiss cheese model" originally proposed by James Reason, U. Manchester, 1990.

Figure A.1 "Swiss cheese" model.

Standby systems with diagnostics. Caution should be exercised when considering standby safety systems (IPLs) that have diagnostics intended to reveal when system failures occur. The temptation is to automatically consider system faults as "revealed" failures with negligible durations. However, diagnostics typically are capable of flagging only 70 to 80% of fail-dangerous failure modes for the IPL system as a whole. IPL system failures should only be considered as revealed failures if they have essentially 100% diagnostics (a rare situation); hence, it is necessary to know exactly what the system diagnostics cover. In addition, even after a system failure is revealed by diagnostics, what is done after the failure is brought to light should also be evaluated. If fault diagnostics warn of a fail-dangerous failure mode but no action is taken to correct the failure, then the failure duration (time required to detect and correct the failure) can be quite long, and the IPL remains unavailable to function on demand.

Quantifying "Double Jeopardy"

To illustrate why "double jeopardy," as properly interpreted, is generally excluded from hazard evaluations and risk analyses, an example situation in which two independent revealed failures occur concurrently will be quantitatively evaluated.

Consider an agitated atmospheric storage tank of polymer that is kept in a molten state for feed to the downstream process by heating with an internal steam coil (see Figure A.2). The polymer has a relatively low melting point, such that its temperature is normally kept at around 30 °C (86 °F). The tank also has a nitrogen blanket for quality purposes.

The scenario that will be evaluated is the concurrent incidence of (1) excess steam to the internal coil, resulting in the polymer being at a hot enough temperature to pose a severe thermal burn hazard, and (2) the nitrogen supply regulator failing open, resulting in the tank being overpressurized. (The consequence of overpressurizing the atmospheric storage tank with polymer at its normal temperature would be a loss event in and of itself, but for purposes of this example is considered much less severe than if the polymer is also heated prior to tank failure.)

This example assumes there is no commonality between the regulator failing open and the temperature control failing high, and that each failure by itself puts the system in an abnormal situation that is readily revealed or detected either by process measurements and other effects on the system as a whole or by the occurrence of a loss event. The following arbitrary failure rate data will be used:

- Temperature control fails high: 0.1/year
- Nitrogen supply regulator fails open: 0.02/year

What is the estimated frequency at which the nitrogen supply regulator fails open and temperature control concurrently fails high? This might be pictured in Fault Tree Analysis format as in Figure A.3. It would <u>not</u> be correct to just multiply the failure frequencies at the AND gate where the two events combined; this should be obvious from the wrong units of measure:

Incorrect: $0.1/\text{year} \cdot 0.02/\text{year} = 0.002 \ (/\text{year})^2$

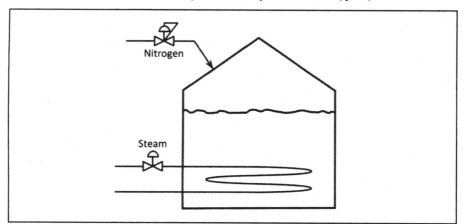

Figure A.2   Simplified example process (agitation and safeguards not shown).

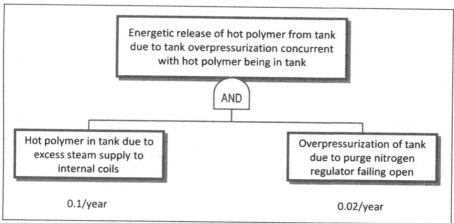

Figure A.3 Abbreviated fault tree for concurrent initiating events example.

Once a failure is revealed, a facility has the opportunity to repair the failed device and return it to a properly functioning state. Hence, it is necessary to have an estimated <u>duration</u> (restoration time) for each event in order to determine the likelihood of occurrence of the concurrent events. For initiating causes, the duration of time before either detecting and correcting the failures or a loss event occurring is typically minimal. For purposes of the continuing example, the following durations will be used:

- N_2 supply regulator fails open: 8 hours to detect, correct and restart
- Excessive steam: 4 hours to detect, correct and re-establish steam flow

The two concurrent, repairable failures are properly considered as two separate scenarios:

1.   The temperature control system fails high first; then, during the 4 h of time before the overtemperature is detected and corrected, the purge nitrogen regulator fails open (and then remains failed open long enough to overpressurize the tank to the point of failure).

2.   The purge nitrogen regulator fails open first; then, during the 8 h of time before the regulator failure is detected and corrected (or the tank is overpressurized to the point of failure), the temperature control system fails high.

These two scenarios are illustrated in Figure A.4, along with the quantification of the events at each stage. At each AND gate, a frequency is properly combined with a dimensionless <u>probability</u> to obtain a frequency of the next higher gate, which are then added together to get the overall frequency of **3 x 10^{-6}/ year**. As expected, this is a very low frequency of having the two failure events happen concurrently.

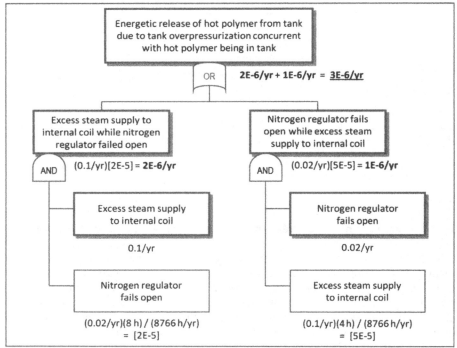

Figure A.4 Quantified fault tree for concurrent initiating events example.

The probability at each branch [set off in square brackets], also termed the *unavailability*, is the fraction of time in the failed state. It is a function of the frequency and duration of one event, combined using the simplified unavailability equation of frequency times duration, which is valid for small products of frequency times duration.

(It is recognized that a proper analysis would need to evaluate how long it would take to heat up the polymer in the tank once excess steam was introduced, and how long it would take to overpressurize the tank once the nitrogen regulator failed open. This analysis only looks at the initiating events, and does not include protective systems such as emergency relief.)

The above example illustrates why, in general, the concurrent incidence of two *revealed* failures has a very low likelihood, and can usually be neglected when there are also scenarios involving unrevealed failures. Two situations where they may not be able to be neglected are:

- When the consequences of the concurrent revealed failures are extremely severe, such that the overall scenario risk is significant.
- When one or more of the total repair times (detection, repair, restart) for the revealed failures cannot be proven and documented to be small.

By contrast, *unrevealed* failures will can occur at any time prior to the revealed-failure initiating event, and the probability the component is in the failed state when an initiating event occurs will generally be much greater than the [2E-5] and [5E-5] unavailability probabilities in the above example. For this reason, when an initiating event is combined with one or more *unrevealed* failures, this is not "double jeopardy." For more details on these calculations, the reader is referred to *Guidelines for Chemical Process Quantitative Risk Analysis* (CCPS 1999).

When the Consideration of Simultaneous Failures IS Valid

In some relatively rare cases, considering "double jeopardy" situations (as properly understood) can be a valid part of a hazard evaluation or risk analysis. These include the following two cases:

- When the combination of the two revealed failures would have an extremely high severity of consequences, such that even, for example, the 3×10^{-6}/ year frequency in the above worked example would be considered significant.
- When the scenario being evaluated involves two revealed failures that may take a non-negligible time to repair.

The following are a few situations where it may not be obvious which system failure is the initiating event and where both failure durations might be non-negligible:

The combination of an indoor hazardous vapor release at a relatively low release rate and the independent failure of a ventilation system designed to reduce the indoor vapor concentration to a nonhazardous level.

The occurrence, at the same time and in the same place, of a flammable vapor or explosible dust atmosphere and an ignition source. This could be either inside a tank, vessel, room or other enclosure, or outside the process in the event of a flammable release.

The combination of a loss of pressure in a utility supply system and the inadvertent opening of an interconnecting valve, thus allowing, for example, backflow of a hazardous material into a water or compressed air system or backflow of air into a nitrogen system.

Note that even in the first two of these three examples, the scenarios could be analyzed by considering one of the two failures as a protective system failure (i.e., the ventilation system for the first example, and ignition source control for the second example). The vast majority of scenarios considered in a hazard evaluation or risk analysis will have one specific initiating event, with all other failures associated with the scenario being either unrevealed (latent) failures or failures such as inadequate alarm response when people or equipment are acting in response to the specific incident sequence.

Incident Example

In a 2008 vessel rupture explosion at the Bayer CropScience facility in Institute, West Virginia, an operational error was made that resulted in the methomyl concentration in the residue treater being much higher than normal. An operator also added hot liquid methyl isobutyl ketone (MIBK) to the residue treater system. The combination of these two factors resulted in an uncontrolled decomposition reaction that was inadequately relieved. Two operators were in the immediate vicinity attempting to correct a perceived vent plugging in response to a high pressure alarm in the unit. (As stated in the investigation report, disabled IPLs also failed to protect against the vessel overpressurization.) These factors might be initially thought of as double jeopardy; however, they are not independent, as the same operating staff was involved in each part of the incident sequence, and neither the high methomyl concentration nor the high MIBK temperature was a revealed failure requiring immediate correction or process shutdown. The CSB investigation report for this incident (CSB 2011) also indicated lack of management enforcement of process safety requirements as a contributing factor to several aspects of the incident. The probability of personnel presence was also not independent of the scenario, since the operators were called out to the process area as a result of a process deviation.

Conclusions

*Most of what is called "double jeopardy" is **not** double jeopardy. Double jeopardy does **not** exist when only one initiating event is present, which is the vast majority of scenarios.*

True "double jeopardy" situations in a hazard evaluation or risk analysis are rare, and most hazard evaluations can be performed without ever mentioning the term. However, scenarios involving concurrent initiating events do exist, and they require careful evaluation when they are encountered if a catastrophic consequence is involved.

Appendix B

Peak Risk Concepts

In Section 2.3, it was mentioned that it may be inappropriate to use time-at-risk enabling conditions when evaluating LOPA scenarios that involve infrequent, short-duration operating modes involving risks with high potential severity. This Appendix discusses why the use of time-at-risk enabling conditions may be inappropriate in such situations, in the context of understanding the nature of "peak risk."

Peak risk can be defined as the level of risk associated with an activity while that activity is occurring (Henselwood 2009). Risks may end up not being adequately controlled if <u>annualized</u> risks are calculated using time at risk as a factor in the risk equation. This can be illustrated in Figure B.1, where certain activities such as startup, shutdown and maintenance activities pose a higher-than-average risk while they are occurring. In this example, the <u>annual-average</u> maintenance risk is quite low for this process unit, taking into account the brief duration of the maintenance activity. However, if the same maintenance personnel are also performing work on several other processes in the same manner, their <u>total</u> risk exposure (*individual risk*) may be inappropriately high.

<u>To illustrate by an extreme example</u>: If a worker is asked to perform an activity that has a 50% chance of fatally injuring the worker, the activity would clearly be considered an intolerably high risk. However, if the worker performs the activity once a year and it takes only one hour to perform, then the annual-average risk calculated using a time-at-risk enabling condition would be

Initiating event frequency = 1 activity per year

Time-at-risk enabling probability = ((1 h duration/yr)/(8766 hr/yr))

Probability of fatality conditional modifier = 0.5

Scenario risk = $1/\text{yr} \times (1/8766) \cdot 0.5 = 6 \times 10^{-5}$ fatalities/yr.

The true risk for this scenario is an average of one fatality every two years, not one every 20,000 years as in the above calculation.

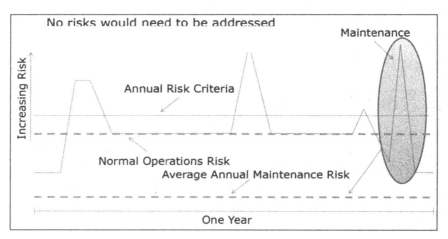

Figure B.1 Illustration of the comparison between annual-average and peak risks. (Credit: NOVA Chemicals. Used with permission.)

Another example would be if LOPAs were conducted on a batch chemical manufacturing facility where different products were made each week of the year. This situation illustrates that peak risk concepts may apply to campaign enabling conditions as well as time-at-risk enabling conditions. If a LOPA was applied to individual products and time at risk was taken into account by using a 1/52 (0.02) factor as an enabling condition, then the calculated risk for each individual product would be only 2% of the total annual risk for the production facility. The campaign enabling condition example of Table 2.2 is reproduced in Table B.1, but with an enabling condition probability of 0.02 instead of the original value of 0.5. The resulting frequency of unmitigated consequence drops to 0.002/year, which would require much less risk mitigation by IPLs or other means to meet a facility's risk criterion than if this enabling condition was not included in the analysis (in effect, recognizing that the risk exposure is continuous by also making other products the remainder of the year).

One approach suggested to address these peak risk concepts is to limit the time-at-risk and campaign enabling condition probabilities to being no lower than 0.1 (Henselwood 2009), as illustrated in Table B.2. This approach effectively limits peak risks to being no greater than ten times the risk level for annual-average risks. This limitation of the peak risks is illustrated in Figure B.2. This provides a balance between allowing time-at-risk enabling conditions and managing the peak risks associated with individual LOPA scenarios. The same peak risk concept can be applied to other high-risk portions such as geographical risks (e.g., not allowing a person to be in a location where the risk exceeds ten times the individual risk criterion, even for short-duration exposures such as during startup of a unit) and one-time activities.

Table B.1 Campaign enabling condition example with a one-week campaign

		Frequency	Probability
...			
Initiating event	Loss of cooling water	0.1/yr	
Enabling condition	Probability that reactor is in condition where runaway reaction can occur on loss of cooling (annual basis)		0.02
Conditional modifiers	None		1
Frequency of unmitigated consequence		0.002/yr	
...			

Scenario title: Cooling water failure results in runaway reaction with potential for reactor overpressure, leakage, rupture, injuries and fatalities. Agitation assumed. Also assumed: Loss of cooling water can be detected before the reactor begins the condition where runaway reaction is possible.

Table B.2 Campaign enabling condition example with a one-week campaign, limiting the time-at-risk probability to a minimum of 0.1

Scenario title: Cooling water failure results in runaway reaction with potential for reactor overpressure, leakage, rupture, injuries and fatalities. Agitation assumed. Also assumed: Loss of cooling water can be detected before the reactor begins the condition where runaway reaction is possible.

		Frequency	Probability
...			
Initiating event	Loss of cooling water	0.1/yr	
Enabling condition	Probability that reactor is in condition where runaway reaction can occur on loss of cooling (annual basis) with time-at-risk enabling condition probability limited to 0.1 or higher		0.1
Conditional modifiers	None		1
Frequency of unmitigated consequence		0.01/yr	
...			

Other approaches include calculating a true annualized risk, as in the example above involving a 50% probability of worker fatality, and calculating a *maximum individual risk,* which is the risk at the location of the most-exposed individual (such as the person working closest to a location where a toxic release or vessel rupture explosion may occur). This would require risk criteria to be available for both societal risk and individual risk, as is the practice in some companies. Societal risk and maximum individual risk are discussed further in *Guidelines for Chemical Process Quantitative Risk Analysis* (CCPS 1999) and in Chastain (2010).

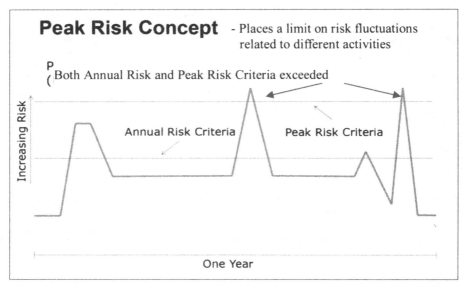

Figure B.2 Concept of placing a limit on how much peak risk can exceed annualized risk. (Credit: NOVA Chemicals. Used with permission.)

One final word of caution is worth emphasizing. As mentioned in Chapter 2, it is not appropriate to use a time-at-risk enabling probability when the initiating cause is activity-based or opportunity-based. Workers are at risk every time they perform the activity or every time they have an opportunity to make a critical error.

Appendix C

Example Rule Set for LOPA Enabling Conditions

As discussed in Chapter 1, Layer of Protection Analysis (LOPA) is a simplified, rule-based approach for assessing scenario risk. An organization employing LOPA will need to establish its own set of rules to be used for conducting LOPAs for its facilities. Such rules can include how enabling conditions and/or conditional modifiers are to be treated. This Appendix presents examples of what those rules might comprise, although a full rule set might also include default values to be used for specific enabling conditions or conditional modifiers, as well as limits on how much risk reduction credit can be taken.

Example Rule Set

Table C.1 gives an example rule set for the use of enabling conditions in LOPAs. The following are examples only. They do not indicate the only possible options, and are not intended to reflect mandatory or universal requirements in any way. For example, some organizations might not limit scenarios to having only one enabling condition, or allow enabling condition and conditional modifier probabilities at a greater resolution than on an order-of-magnitude basis. A rule set for conditional modifiers would have requirements similar to those listed here for enabling conditions, but with a different definition and possibly a different restriction on the number of conditional modifiers that can be employed in analysis of a given scenario.

Table C.1 Example rule set for use of enabling conditions in LOPAs *(see text)*

Enabling condition rule	**Comments, examples** (see Notes at end of table)
1. The definition of an enabling condition must be met; i.e., a condition that is not a failure, error or a protection layer but makes it possible for an incident sequence to proceed to a consequence of concern.	An enabling condition consists of an operating phase or condition that does not directly cause the scenario being evaluated, but that must be present or active in order for the scenario to proceed to a loss event. Enabling conditions are expressed as dimensionless probabilities.
2. Only one enabling condition is allowed per scenario. *(Note 1)*	This reduces the likelihood of double-counting enabling conditions such as time at risk and campaign factors.
3. Enabling conditions are limited to order-of-magnitude probabilities. *(Note 2)*	A probability of 0.2 or less can be rounded down to 0.1. A probability greater than 0.2 must be rounded up to 1. Similar rounding is to be used for lower values; e.g., 0.03 is rounded up to 0.1. *(Note 3)*
4. Enabling condition probabilities will typically be **1**, **0.1** or **0.01**.	Use of an enabling condition probability less than 0.01 requires a documented, approved exception. *(Note 4)*
5. The enabling condition must not be correlated with the scenario initiating event or with any independent protection layers (IPLs).	Typically, this means there are no features or attributes shared with the initiating event and the IPLs, and a human (or group of humans) is only used once in an incident scenario. *(Note 5)* Also, no utility failure will cause the enabling condition to be true while also initiating the scenario or defeating any IPL. This rule eliminates the need to do a common cause or common mode failure analysis, thereby greatly simplifying the analysis.

Table C.1 *(continued)*

Enabling condition rule	Comments, examples (see Notes at end of table)
6. The enabling condition must represent a true risk-reduction factor that can be expected to be valid over time. *(Note 7)*	If the enabling condition is included in a LOPA for an existing facility, it must be supported by actual plant or historical data. *(Note 6)* If the enabling condition is included in a LOPA for a new design, it must be validated within one year after commissioning of the new facility by comparing assumptions to actual practice. *(Note 7)* Facility changes that might impact enabling condition probabilities must be evaluated as to the potential safety and health impact of such changes as part of the faciility's management of change procedures. *(Note 7)* Revalidation of enabling condition probabilities is to be included as part of each LOPA update. *(Note 7)*

Notes:

1. It is uncommon but possible to have more than one enabling condition in a LOPA. More commonly, there may be two or more factors that would be combined to give, for example, a time-at-risk enabling condition probability. Some organizations might not have any restrictions on the number of enabling conditions for a single scenario.

2. Other options include restricting enabling conditions to half orders of magnitude resolution or to one significant figure.

3. Other rounding schemes are also possible, such as rounding to the nearest half order of magnitude or having a probability of 0.3 or less be rounded down to 0.1. The most conservative approach would be to always round up to the next highest order of magnitude. For example, a calculated enabling condition probability of 0.015 would have to be rounded up to 0.1 using this approach.

4. Some organizations may draw the line differently for this rule, such as not allowing an enabling condition probability less than 0.1 or less than 0.001.

5. This is more likely to be an issue with conditional modifiers than with enabling conditions.

6. Some organizations may express this rule differently; for example, they may use enabling conditions not supported by actual data but use standard values for particular enabling conditions or use input from experienced operational personnel.

7. This is likely to be approached differently by various organizations, depending on the reason for conducting the LOPA and whether there is any requirement for updating LOPAs on a periodic basis, since such a requirement is not mandated as it is for PHAs.

References

API RP 579. Fitness-for-Service, American Petroleum Institute, Washington, DC.

API Std 530. Calculation of Heater-Tube Thickness in Petroleum Refineries, American Petroleum Institute, Washington, DC.

CCPS 1995. Center for Chemical Process Safety, *Guidelines for Consequence Analysis of Chemical Releases,* ISBN 978-0-8169-0786-1, American Institute of Chemical Engineers, New York.

CCPS 1996. Center for Chemical Process Safety, *Guidelines for Use of Vapor Cloud Dispersion Models, 2nd Edition,* ISBN 978-0-8169-0702-1, American Institute of Chemical Engineers, New York.

CCPS 1999. Center for Chemical Process Safety, *Guidelines for Chemical Process Quantitative Risk Analysis, 2nd Edition,* ISBN 978-0-8169-0720-5, American Institute of Chemical Engineers, New York.

CCPS 2001. Center for Chemical Process Safety, *Layer of Protection Analysis: Simplified Process Risk Assessment,* ISBN 978-0-8169-0811-0, American Institute of Chemical Engineers, New York.

CCPS 2008. Center for Chemical Process Safety, *Guidelines for Hazard Evaluation Procedures, Third Edition,* ISBN 978-0-470-25147-8, American Institute of Chemical Engineers, New York.

CCPS 2009. Center for Chemical Process Safety, *Guidelines for Developing Quantitative Safety Risk Criteria,* ISBN 978-0-470-26140-8, American Institute of Chemical Engineers, New York.

CCPS 2012a. Center for Chemical Process Safety, *Guidelines for Engineering Design for Process Safety, 2nd Edition*, ISBN 978-0-470-76772-6, American Institute of Chemical Engineers, New York.

CCPS 2012b. Center for Chemical Process Safety, *Guidelines for Evaluating Process Plant Buildings for External Explosions, Fires, and Toxic Releases, 2nd Edition,* ISBN 978-0-470-64367-9, American Institute of Chemical Engineers, New York.

Chastain 2010.   J. W. Chastain, "Ensuring Consistency of Corporate Risk Criteria," 6th Global Congress on Process Safety, San Antonio, Texas, March.

CSB 2002.   Investigation Report, Refinery Incident, Report No. 2001-05-I-DE, U.S. Chemical Safety and Hazard Investigation Board, Washington, DC, October.

CSB 2009.   Investigation Report, Sugar Dust Explosion and Fire, Report No. 2008-05-I-GA, U.S. Chemical Safety and Hazard Investigation Board, Washington, DC, September.

CSB 2011.   Investigation Report, Pesticide Chemical Runaway Reaction Pressure Vessel Explosion, Report No. 2008-08-I-WV, U.S. Chemical Safety and Hazard Investigation Board, Washington, DC, January. DoD 2007.   U.S. Department of Defense, Unified Facilities Criteria (UFC), DoD Minimum Antiterrorism Standards for Buildings, UFC 4-010-01 including change 1, 22 January.

EPA 2012.   U.S. Environmental Protection Agency, Office of Emergency Management, Aerial Location of Hazardous Atmospheres (ALOHA), http://www.epa.gov/emergencies/content/cameo/aloha.htm.

Gubinelli et al. 2004.   G. Gubinelli, S. Zanelli and V. Cozzani, "A simplified model for the assessment of the impact probability of fragments," *Journal of Hazardous Materials A116*, 175-187.

Henselwood 2009.   F. Henselwood, "The Application of Peak Risk Concepts within a Corporate Risk Program," 5th Global Congress on Process Safety, AIChE Spring National Meeting, Tampa, Florida, April.

Johnson 2010.   R. W. Johnson, "Beyond-Compliance Uses of HAZOP/LOPA Studies," *Journal of Loss Prevention in the Process Industries 23*(6), November, 727-733.

Johnson 2011. R. W. Johnson, *HAZOP Studies*, AIChE eLearning Course ELS104, New York: American Institute of Chemical Engineers, aiche.learn.com or www.aiche.org/resources/education/elearning.

HSE 2011.   UK Health & Safety Executive, "Assessment of the Dangerous Toxic Load (DTL) for Specified Level of Toxicity (SLOT) and Significant Likelihood of Death (SLOD)," www.hse.gov.uk/chemicals/haztox.htm.

Moosemiller 2009.   "Development of Algorithms for Predicting Ignition Probabilities and Explosion Frequencies," *5th Global Congress on Process Safety,* New York: American Institute of Chemical Engineers, April.

NFPA 654.   Standard for the Prevention of Fire and Dust Explosions from the Manufacturing, Processing, and Handling of Combustible Particulate Solids, National Fire Protection Association, Quincy, Massachusetts, www.nfpa.org.

Pitblado and Tahilramani 2009.   R. M. Pitblado and R. Tahilramani, "Risk Communications: Web-sites for Barrier Diagrams and Process Safety," Mary Kay O'Connor Process Safety Conference, College Station, Texas, October.

Stephens 1970.   M. M. Stephens, Minimizing Damage to Refineries, U.S. Department of the Interior, Office of Oil and Gas, February.

Summers and Hearn 2011.   A. E. Summers and W. H. Hearn, "Risk Criteria, Protection Layers and Conditional Modifiers," 3rd CCPS Latin American Process Safety Conference and Expo, Buenos Aires, Argentina, August.

Swiss Re 1998.   ExTool User Manual, Swiss Reinsurance Company, Zurich, Switzerland.

Index